The Value of Food

THE VALUE OF FOOD

PATTY FISHER, B.Sc. (H & SS), F.R.S.H.
Formerly Lecturer in Nutrition, Bath College of Higher Education

and

ARNOLD E. BENDER,
B.Sc., Ph.D., F.R.S.H., F.I.F.S.T.
Professor of Nutrition, and Head of Department of Food Science and Nutrition, Queen Elizabeth College, University of London

THIRD EDITION

OXFORD UNIVERSITY PRESS
1979

Oxford University Press, Walton Street, Oxford OX2 6DP
OXFORD LONDON GLASGOW
NEW YORK TORONTO MELBOURNE WELLINGTON
KUALA LUMPUR SINGAPORE JAKARTA HONG KONG TOKYO
DELHI BOMBAY CALCUTTA MADRAS KARACHI
IBADAN NAIROBI DAR ES SALAAM CAPE TOWN

First edition 1970
Second edition 1975
Third edition 1979

British Library Cataloguing in Publication Data

Fisher, Patty
The value of food. – 3rd ed.
1. Nutrition
I. Title II. Bender, Arnold Eric
641.1 TX353 79–41090

ISBN 0–19–859465–8
ISBN 0–19–859478–X Pbk

PRINTED PHOTOLITHO IN GREAT BRITAIN
BY EBENEZER BAYLIS AND SON LTD.
THE TRINITY PRESS, WORCESTER, AND LONDON

Contents

Acknowledgements

The following diagrams have been reproduced by kind permission; Table 5, abbreviated from Robertson and Reid, *Lancet* 1952, i, 940; Table 36, A. M. Brown, *Nutrition*, 1962, 16, 117; Fig. 14, Harries and Hollingsworth *British Medical Journal*, 1953, i, 75; Figs. 15, 16 and 17 from the film *Protein and Health* presented by the Flour Advisory Bureau; Figs. 21, 24, 25, *Encouraging the use of Protein Foods* (Food and Agriculture Organization).

Preface to third edition

The subject of nutrition keeps growing, not only as a result of the increase of our knowledge of the subject but because students of nutrition must now for many reasons make themselves familiar with a number of topics once regarded as being on the fringes of the field. These include the ever-growing area of food legislation, which affects food composition and nutritional value, as well as labelling and food additives and their safety.

With the growing interest of consumers and the advent of nutritional labelling elsewhere, there is greater interest in topics such as enrichment of foods and changes taking place during processing—hence these sections have been expanded.

The area of social nutrition—why people eat what they do—is of such obvious importance that it must be touched on, even if lack of space means that it cannot be dealt with in any detail.

There is a growing realization that some of the disorders common in industrialized communities, the so-called diseases of affluence, are related to diet. Novel foods are being produced that may, in time, greatly affect the farmer, the caterer, the cook, and the nutritionist.

So what set out originally to be a short, simple, concise book has necessarily grown longer. The aim is still to provide a fairly comprehensive coverage, leaving it to teachers and students to select what they need.

Extensive revision of many of the tables has been necessary since the values in the new edition of McCance and Widdowson's *The Composition of Foods* by A. A. Paul and D. A. T. Southgate (H.M.S.O.) differ in many respects from the original values. Even where the differences are small, the reader may otherwise have difficulty in comparing his or her own calculations with those given here.

Most values quoted are taken from that publication, with a limited number from *The Composition of Tropical Foods* by B. S. Platt (H.M.S.O. 1945) and from the *Manual of Nutrition* (H.M.S.O. 1978).

P.F.
A.E.B.

Preface to second edition

Apart from the continuous changes that result from research findings there have been several changes in nutritional thinking over the past few years, and some measure of agreement has been reached on topics such as nomenclature of vitamins and figures for recommended intakes of nutrients.

In particular, the energy expenditure of the body and the energy content of foods are now expressed in joules, and quantities of all vitamins are expressed in micrograms. As well as including the new figures for recommended intakes and for calculating vitamin A in terms of retinol equivalents this new edition brings up to date the various food regulations.

Since some official bodies and certainly many textbooks will continue to use the old nomenclature (including calories) for some time to come, in this book quantities are given in both sets of units.

1975

P.F.
A.E.B.

Preface to first edition

The nutritional aspects of the food that we eat and the practical applications of the science of nutrition are subjects of growing interest and importance. The food field is a very broad one, extending from the sea and the farm at one end, through storage, processing, and preparation, to the physiological and biochemical processes of the living body at the other end.

This book attempts to provide enough information on these topics to satisfy the needs of the many students in the various training courses that include a study of food. Some courses might not call for as much theory as is provided; rather they may require the applications, they may need the verdict, others may need the evidence. By providing both in the same book we hope to satisfy the requirements at several levels—domestic science at school and college, catering training, and the various food crafts.

Some of the foods mentioned may appear a little unusual among the cabbages and potatoes but they have been included so that the book may possibly be of some use abroad and to the foreign students being trained in this country, apart from the fact that many hitherto unusual foods are becoming almost commonplace.

For the purpose of compiling menus we have used F.A.O. figures where they have been established; in other cases U.S.A. figures have been used, since the new U.K. figures were not available at the time.

The Value of Food aims at a comprehensive coverage leaving it to the students and teachers to select what they need.

P.F.
A.E.B., 1969

A note on units

There are many differences internationally in the units used—calories are used by some, joules by others, and both side by side by yet others. Ounces and grams are both still in use, as well as international units of some vitamins in some countries. In the present state the nutritionist has no choice but to be familiar with all systems, so in most cases old and new units are stated side by side. Apparent discrepancies arise by conversion followed by rounding off the figures if the original measurements were made in old or in new units. However, these are not serious and the nutrient composition of foods varies considerably from sample to sample largely because of their natural variation.

1 Food for thought

Good health depends on many factors—food, heredity, climate, hygiene, exercise—and food is the most important of all these. This fact was driven home in an experiment carried out in India in the 1920s by Sir Robert McCarrison.

India is a land of many races and the cities are filled with people of a surprising variety of physique and fitness. There are the tall, healthy-looking Sikhs, Hunzas and Pathans from the north. There are people who are small and thin and others who are large and fat, neither looking particularly healthy—they come from the south. Then there are the small, well-built, hardy Gurkhas from the mountains of Nepal.

What is the reason for these differences? Is it race, living conditions or climate? Sir Robert McCarrison designed an experiment to find out.

He took six pairs of rats from the same family, of the same age and with the same standards of health, and kept them in the same climate with the same standards of hygiene. The only way they differed was in their diet—each pair was fed on one of the diets from the different races of India. One pair had white rice seasoned with curry and garlic, with occasional eggs and fish, but no milk or vegetables—the diet of south India. Another pair had the same with the addition of fats, spices, dried fish and sugar. Yet another had the diet of north India—chapattis (pancakes made from brown flour) with fruit and vegetables, both raw and cooked, lentils and peas, with occasional milk and ghee (a kind of butter).

After several months on these different diets the six pairs of identical rats had become six groups of quite different rats. Some were large and fat with grey fur moulting in patches, others were thin and weak. Both these groups were short-lived and had few offspring. Another group were splendid specimens with sleek fur and numerous progeny. These had been fed on the diets of north India.

All the rats had the same heredity and lived in the same climate with the same standards of hygiene; so clearly, food was the deciding factor in controlling their health, stamina and physique.

MALNUTRITION

In many countries malnutrition is largely responsible for the high death-rate among children. The death-rate of children under 1 year of age in some of the poorly-fed countries may be 10 times as great as that in the well-fed countries, and the death-rate of children between 1 and 4 years old may be 40 times as great. The chief causes of this high death-rate are disease and malnutrition, the one precipitating and perpetuating the other.

The Japanese have shown what improvements can be achieved by diet; they used to be short in stature and lived largely on rice and a little fish. Since 1946 the children have had school meals of bread, milk,

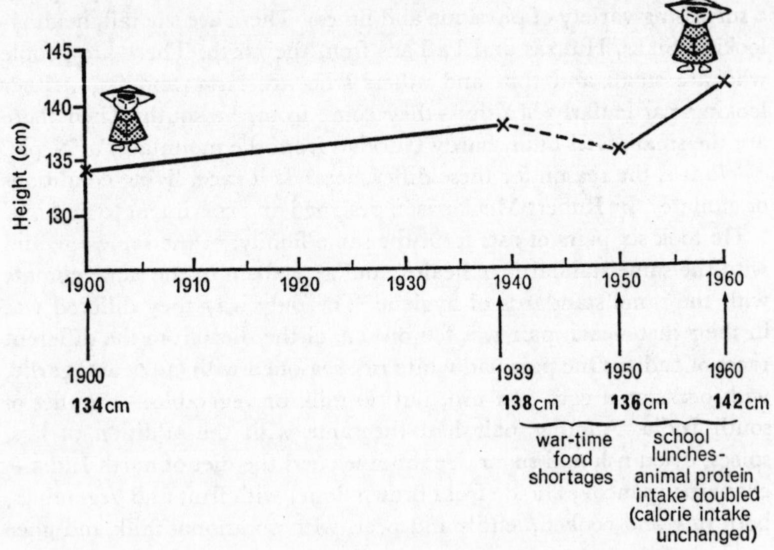

Fig. 1 Height of Japanese children at 12 years of age

cheese and vitamin C-rich fruit. Now they are taller and heavier than their parents were at the same age and they need larger desks at school (see Fig. 1).

GREAT BRITAIN

In Great Britain, too, records show that as diets have improved children have grown taller and heavier than the previous generations (see Fig. 2). Of course there have been other changes in living standards at the same time, such as improvements in hygiene and surroundings, but food remains the decisive factor. Events at Stockton-on-Tees in the years 1923–32 illustrate this (see Fig. 3).

The slum area of one portion of the town was due for rehousing in

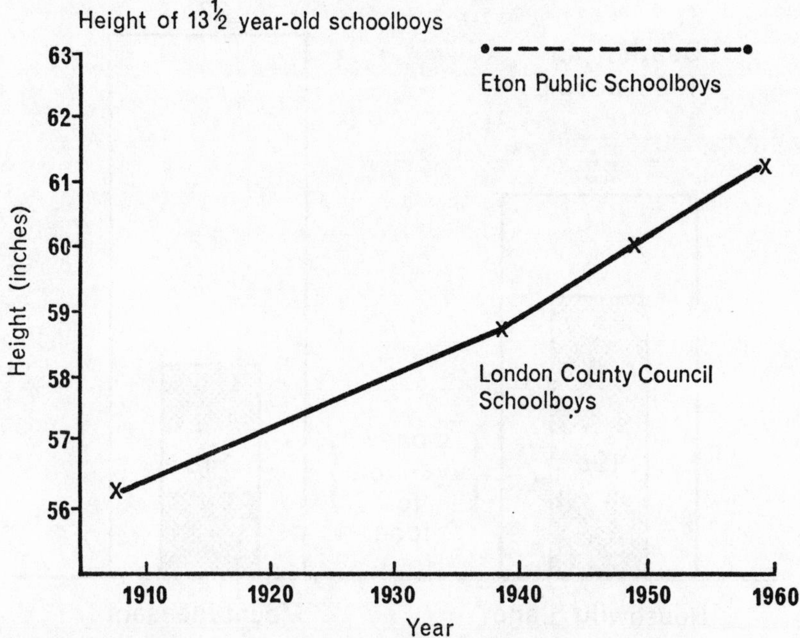

Fig. 2 Relation between standard of nutrition and height of children

1923 and the people from Housewife Lane, 700 of them, were moved into new self-contained houses in Mount Pleasant. In the adjacent Riverside area 1 300 people stayed for the time being in the slums. The new houses had baths, a kitchen range, a well-ventilated store and a wash-boiler—all that one would hope for to make a family happy and healthy.

Before the removal the death-rate in both these slum areas was greater than in the rest of the town: 23 per 1 000 in Housewife Lane and 26 per 1 000 in Riverside. The removal into the new housing estate with all its advantages would be expected to bring about an improvement in health which would be reflected in a fall in the death-rate. But, when the figures were collected 5 years after rehousing, the death-rate had *risen* in the new Mount Pleasant estate from 23 to 33 per 1 000, while the people in the old slum in Riverside showed a slight improvement to 23.

This surprising result needs some explanation. Poor families in slum conditions can be expected to have poorer health than those living under better conditions, not the reverse. The explanation was found in their diet.

Very many families in both Riverside and Mount Pleasant were unemployed and had only a small, fixed sum for buying food. When

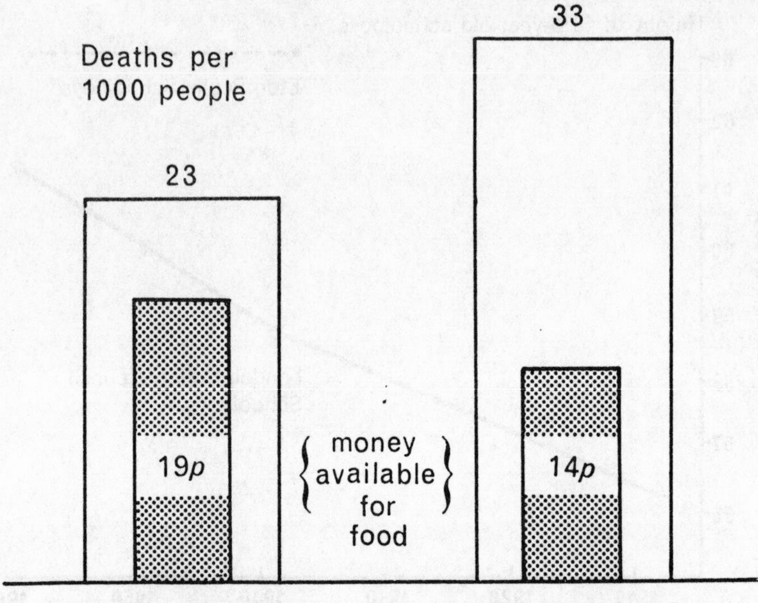

Fig. 3 Relation between money spent on food and health

they were moved into the new houses the rents were raised from 23p per week (in 1927) to 45p per week, and so less was available for food. In fact their old neighbours left in the slums were able to continue to spend the sum of 19p per head per week on food, while the Mount Pleasant inhabitants in their new houses had only 14p per week for food. The effects of the improved housing and sanitation were overcome by a poorer diet.

ARE WE PROPERLY FED?

Severe malnutrition is fairly obvious since it gives rise to defined diseases such as beri-beri, pellagra, scurvy, and so on, but mild degrees of undernutrition are difficult and sometimes impossible to diagnose. It is very rare indeed to find severe malnutrition in the developed countries of the world, but at the same time it is difficult to decide whether everyone is adequately fed.

An adequate diet will, by definition, provide full stores in the body of all the materials required for growth and repair of tissues, and the subject will be in good health when examined clinically:

Adequate diet : Full body stores : Good health

What would happen if the intake of one or more nutrients fell slightly below a person's needs? The first stage would be that the body stores would be reduced, and would fall continually, until eventually there might be no stores at all. This would have no effect on the health of the person and could only be revealed if samples of the blood or tissues were analysed for their stores of nutrients, i.e. by biochemical examination. In other words this means that before it is possible to draw conclusions about the nutritional status of an individual it is necessary to carry out a biochemical as well as a dietary assessment:

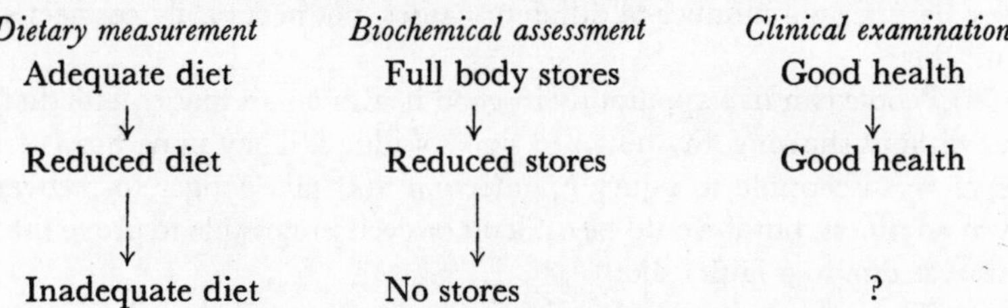

When the diet is even more deficient, or the slight deficiency has continued over a long period, the normal functioning of the organs and tissues may be impaired, but there may be no visible sign of ill-health. A disturbance of function can be shown by biochemical measurements. Since there are no clinical signs of ill-health the condition is called 'subclinical malnutrition'.

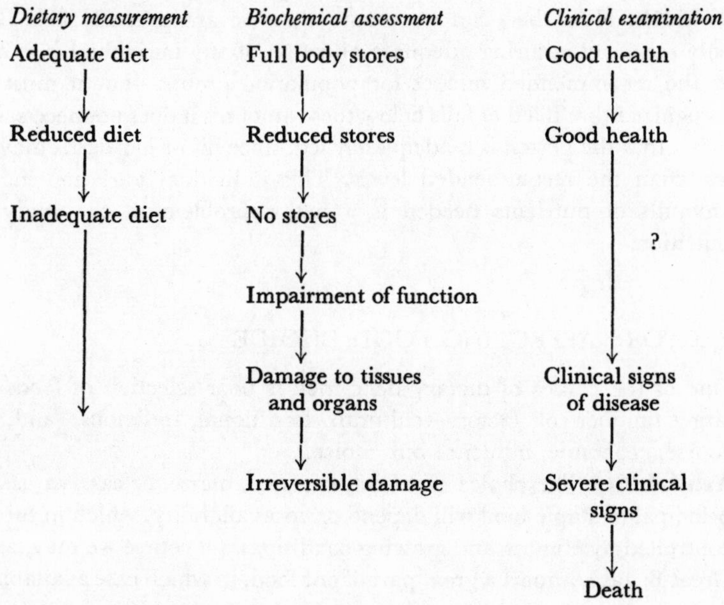

It is only when the tissues are actually damaged that clinical signs appear. This illustrates three things.

(1) Assessment of the nutritional status of an individual can be made only after three types of examination have been made, dietary, biochemical, and clinical. Any one of these will *not* provide the information required, because people differ a great deal in their individual need for the various nutrients, and also in their normal biochemical levels of substances like mineral salts and vitamins and co-enzymes made from vitamins. Clinical signs, even when severe, are not always specific, but may be due to a number of different causes, not necessarily connected with diet.

(2) People can live apparently in good health on an inadequate diet, i.e. without showing any outward signs of illness. They may, however, be more susceptible to injury or infection and take longer to recover from an illness, but it would be difficult or even impossible to prove that this was due to a faulty diet.

(3) The only early warning sign of malnutrition is that the person is known to be getting less in his diet than he needs. Unfortunately we do not know the needs of individuals (which vary greatly) but only the average needs of population groups, such as adult men or women, or children of various ages. It is these averages plus an extra to cover the higher needs of some individuals that are the basis of tables of recommended daily intakes as listed on pp. 184–6—they do *not* apply to individuals. How then can we ensure that we are properly fed? The only means of ensuring adequate nutrition of any individual is to aim at the recommended intakes for population groups—but it must be recognized that if a diet falls below these amounts it does not necessarily mean that the person is inadequately fed, since his or her needs may be less than the recommended levels. This individual variation in the amounts of nutrients needed is a major problem in the study of nutrition.

FACTORS AFFECTING FOOD CHOICE

One of the causes of dietary deficiency is poor selection of foods. A large number of factors—cultural, traditional, religious, and, of course, economic, influence our choice.

Availability The choice of wheat, rye, rice, maize, or cassava as our principal or staple food will depend on its availability, which in turn is controlled by climate and growing conditions. Of course we may, as in Great Britain, import a great part of our food, in which case availability depends on national wealth and ability to compete in the world food markets.

Social factors Social changes, such as housewives going out to work, influence our dependence on ready-made convenience foods and take-away meals. The introduction of Flexitime (flexible working hours in offices and factories) affects the time available to housewives for cooking. Distance from school or work may make us eat away from home at lunch time.

Fashion From time to time people go in for vegetarianism, health foods or 'natural foods free from additives'; or there may be a rise in the popularity of kippers or coffee, for a variety of reasons.

New foods, affluence, and **advertising** all combine to alter our eating habits. The increasing use of the domestic freezer and, more recently, the microwave oven, can alter our eating habits.

Religion, tradition, taboos, and **cultural patterns.** These may cause us to refrain from certain foods completely, permanently, or at certain times of the year, or to indulge in special dishes. They may explain why dogs and cats are not eaten in Europe, why horse meat is not eaten in Britain, although it is relished in France, and why black puddings and tripe are commoner in the north of England than the south.

Price This can be controlled by taxes and subsidies as well as good and bad harvests, and obviously plays an important role in our food selection and therefore in our nutritional health.

The whole subject of food habits is clearly an important part of the study of nutrition. It is interesting to trace from one's personal experiences and that of one's friends what factors have caused a change, if any, in one's diet. Was it foreign travel, advertising, a knowledge of nutrition, or having the money to eat at a restaurant that introduced a new food into our diet?

GOOD NUTRITION, BETTER HEALTH

A study of nutrition can tell us the foods most suitable for health. As a prosperous nation we have enough money to buy them, but many people still make a poor choice of foods. What foods are needed for good health? Foods are needed for three main purposes: muscular work, vitality, and body construction and repair. And since we eat some foods purely for pleasure, we can add a fourth category—FUN.

2 Foods for muscular work

Working muscles, like the engine of a car, need both fuel and a spark to release the energy from that fuel. The fuel is of two kinds, carbohydrates (i.e. starches and sugars) and fats. To release the energy from the fuel requires the B vitamins. Just as an engine cannot get the energy from petrol without a sparking plug, so the body cannot get the energy from food without the B vitamins.

It is easy to make sure that we eat enough energy-foods for fuel. So long as food is available we always eat enough to satisfy hunger and that means satisfying our needs for fuel. It is not so easy to make sure that we have enough of the B vitamins. We have no instinct to choose the right foods and by choosing the wrong ones it is possible to satisfy our hunger but still be short of vitamins. So we do need to have some knowledge of nutritional values.

Part 1 THE FUEL

CARBOHYDRATES

We eat a very large number of different foods but they all contain only five sorts of *nutrients*—carbohydrates, fats, proteins, mineral salts and vitamins. The carbohydrates and the fats are energy-foods. Carbohydrates provide most of the energy in almost all human diets—in some areas as much as 80%.

Chemically, carbohydrates are composed of carbon, hydrogen and oxygen, with 2 atoms of hydrogen for each one of oxygen. The group includes the sugars (mono, di, and trisaccharides) and starch, cellulose, and glycogen, which are polysaccharides (see Fig. 4).

Sugars The simplest sugar and the one that is used directly by the muscles is glucose or dextrose. It is found in fruits and is also called grape sugar. All the carbohydrates that we eat and absorb are eventually converted into glucose and so the blood sugar, the form of carbohydrate that circulates in the blood stream to all parts of the body, is glucose, $C_6H_{12}O_6$.

Starch is composed of large numbers of glucose units linked together, hence it is a polysaccharide. When it is broken down (or hydrolysed), either during digestion or by heating with acid, it first breaks down into

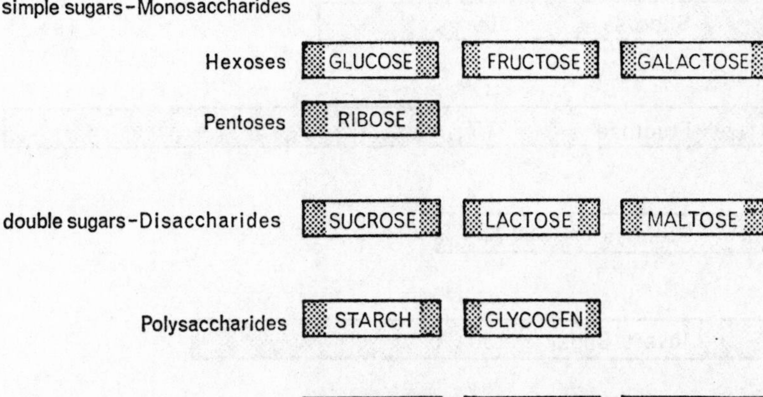

Fig. 4 Carbohydrates

smaller groups of glucose units called dextrins, and then into glucose itself. Partially hydrolysed starch, consisting of a mixture of dextrins, maltose and glucose, is one of the principal ingredients of sugar confectionery and is called confectioners' glucose (or glucose syrup or corn syrup). Whereas chemists tend to use the name glucose for the pure sugar, the confectionery trade uses this name for glucose syrup and calls the pure sugar dextrose.

Pure glucose or dextrose is manufactured commercially by the complete hydrolysis of starch, usually maize (or corn) although any form of starch, such as wheat, rice or potato starch, may be used. As glucose is not as sweet as cane sugar it is sometimes added to foods and drinks—so called glucose-enriched—to increase their energy content without making them unbearably sweet (see Fig. 5).

There is a second simple sugar or monosaccharide called fructose or fruit sugar, which occurs together with glucose in many fruits. In the body it is converted into glucose. Fructose is tolerated by diabetics and often used in formulating diabetic foods.

There are a number of double sugars or disaccharides, so called because they consist of two monosaccharides linked together. The commonest is sucrose, cane (or beet) sugar—which is a double molecule of glucose and fructose and is hydrolysed to these simple sugars during digestion.

There are two other disaccharides commonly found in our food: (1) lactose or milk sugar, which is broken down during digestion to glucose and another simple sugar, galactose, and (2) maltose or malt sugar, which is broken down to two molecules of glucose.

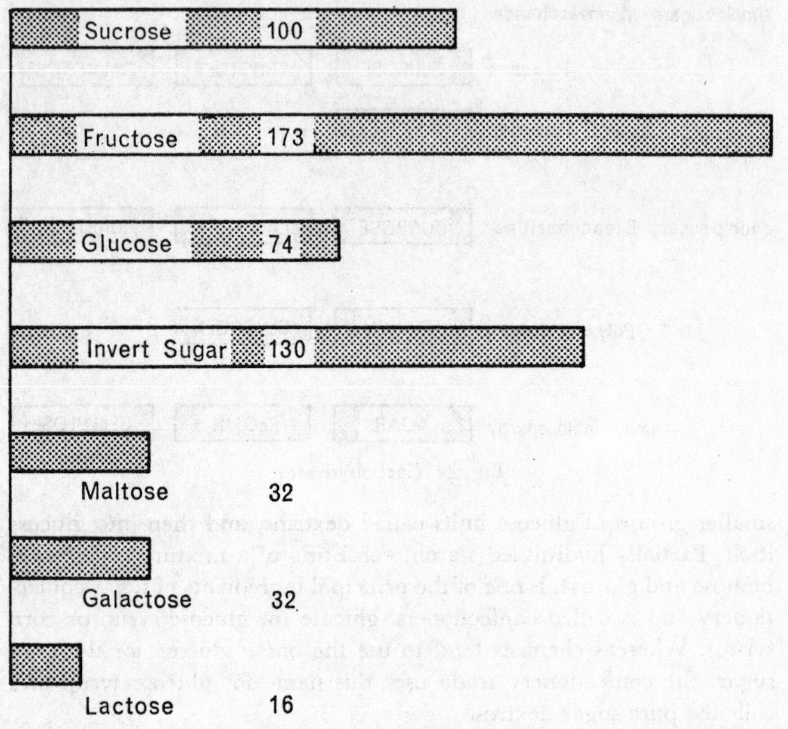

Fig. 5 Relative sweetness of the sugars

Common carbohydrates As we have seen, glucose is found in nature both free and in complexes made up of large numbers of glucose units such as starch, glycogen and cellulose.

Starch is found in wheat and other cereals, in roots such as cassava, yam, arrowroot, in storage organs such as potatoes and, in smaller quantities, in many other plants. As these foods form the major part of man's diet, starch is our major energy-food.

Only plants form starch. During digestion in the animal body it is hydrolysed to glucose so there is no starch in animals. Instead there is another glucose complex called glycogen, and this is the form in which the body stores glucose in the liver and muscles.

The third complex of glucose is cellulose but this is not digested at all by human beings and passes through the body unchanged. Ruminants such as the cow and sheep can use cellulose as a source of energy only because the bacteria in their rumen break it down.

There are other carbohydrates which are not sources of energy but have culinary uses. Agar, a carbohydrate extracted from seaweed, is used in ice-creams and jellies. Pectin, extracted from fruits such as apples and from citrus peel, is used to aid the setting of jam, and in confectionery.

These carbohydrates are not digested and therefore can safely be eaten by diabetics. There is another sugar derivative, sorbitol, which can be tolerated by diabetics because it is rather slowly absorbed although it does serve as a source of energy. It is rather expensive and is used only in diabetic foods.

Chemical derivatives of cellulose such as methyl cellulose which, like cellulose itself, are not digested and so have no food value, are used in synthetic creams and 'slimming' foods.

The artificial sweetening agents saccharine and cyclamate are not carbohydrates; they have no nutritive value and are simply flavouring.

FATS

The other group of energy foods comprises the fats and oils, often referred to as 'lipids'. Oils are merely fats which are liquid at room temperature because they contain more of the unsaturated fatty acids, as described below.

Fats are a much more concentrated fuel than carbohydrates or proteins since they supply 9 kcal (37 kJ) per g compared with 4 kcal (17 kJ).

In most industrialized countries fats supply as much as 40–45% of the total energy intake and have been blamed for some of our 'modern' diseases. In many poor countries they supply as little as 10% of the energy (because they are expensive) and this is sometimes the reason why total energy intake is inadequate. With very little fat in the diet a much larger volume of food is needed. Another problem of a low-fat diet is that the vitamin A in particular, and other fat-soluble vitamins, are very poorly absorbed.

Fats are also of special nutritional value since they carry in solution the fat-soluble vitamins (A, D, E, and K). Milk fat, and therefore butter and cream, contain vitamin A and some vitamin D; fish-liver oils are very rich in these two vitamins; red palm oil is extremely rich in carotene; margarine is made from oils that have no vitamins A and D and these are added during manufacture; many vegetable oils, such as cotton seed, peanut, and especially wheat-germ contain vitamin E.

Among their other attributes, fats carry much of the flavour of foods, lubricate foods such as bread, and help swallowing; and, of course, they add variety to methods of cooking.

So far as the physiology of the body is concerned fats provide an economical way of storing energy. All energy consumed surplus to immediate needs, whether as carbohydrate, protein, or fat, is stored in the body as fat, since there is only a limited capacity to store carbohydrate (as glycogen). People can store anything from 10 to 100 kg of fat, enough to last from between a month and a year as reserve stores!

The layer of fat stored under the skin also acts as a heat insulator and in aquatic animals helps to keep them afloat.

TYPES OF FATS

There are very many different types of fats in our diet, ranging from hard ones like lard, to liquid oils like soybean and olive, and including fatty foods such as cream, kippers, and sardines. They are, however, similar in their chemical structure since they consist mainly of glycerol combined with fatty acids.

Glycerol is a sweet, colourless liquid popularly called glycerine. It is made of three carbon atoms each carrying one hydroxyl group. Each hydroxyl group can combine with a fatty acid, so fats are mainly triglycerides (Fig. 6). All three fatty acids can be the same or any of a dozen or more different ones; this is why there are so many different fats.

Triglycerides comprise by far the greater part of fats and oils, the remainder being special types of fats, such as cholesterol, and complexes containing phosphate (phospholipids, such as lecithin).

Fatty acids are of two main types: (a) saturated, meaning that all the carbon atoms in the fatty acid carry their full complement of hydrogen atoms (examples are butyric acid with 4 carbons, palmitic with 16, and stearic with 18) and (b) unsaturated ones, where some of the hydrogen atoms are missing and they have double bonds

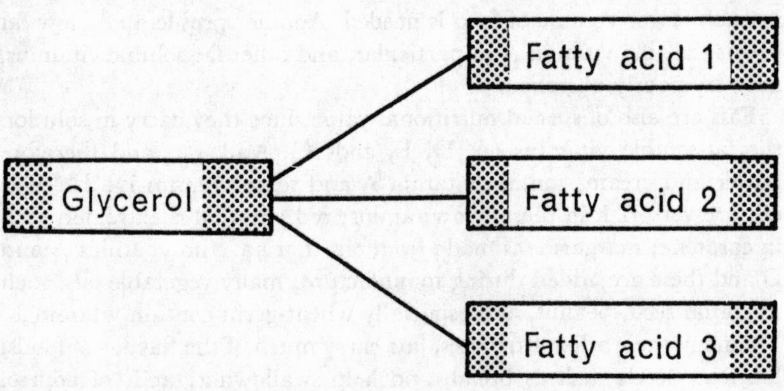

Fig. 6 Triglycerides

between the carbon atoms instead. Examples are oleic acid with 18 carbon atoms and one double bond, erucic acid with 22 C and 1 double bond, linoleic acid, 18 C and 2 double bonds, linolenic acid, 18 C and 3 double bonds, and arachidonic acid, 20 C and 4 double bonds.

The reason why vegetable oils are liquid is that they contain a high proportion of unsaturated fatty acids, chiefly oleic acid, compared with harder fats. As explained below it is possible to restore the missing hydrogen atoms (hydrogenation) and convert liquid oils into fats of different degrees of consistency for shortening and for margarines.

Fatty acids with several double bonds are termed 'polyunsaturated fatty acids' (PUFA), and include linoleic, linolenic, and arachidonic acids (this last one is found only in animal fats in small amounts).

Linoleic acid is a dietary essential and, indeed, was once called vitamin F. However, it is required in very small amounts of about 1 g per day and it is only under very exceptional circumstances that any deficiency has ever been observed in human beings.

The body can make linolenic and arachidonic acids from linoleic acid, so the last is the essential one, although people sometimes refer to all three as 'essential fatty acids'.

However, PUFA (the three listed above and a few others) play a part in controlling, to some extent, the amount of various types of fats found in the blood-stream, as discussed under the section on heart disease, (p. 101). Certain saturated fatty acids raise blood fats (including blood cholesterol), and PUFA reduce them. Hence in attempting to affect coronary heart disease it is often advised to increase PUFA and decrease saturated fats, which will alter the polyunsaturated to saturated (**P/S**) ratio.

RANCIDITY

All fats contain some unsaturated fatty acids and can pick up oxygen at the double bond and so go rancid. This oxidation is accelerated by light, heat, and traces of metals, especially copper, and this is why fats should be stored in the dark under cool conditions. (When potato crisps were marketed in clear plastic packets they went rancid in the shop window in as little as 24 hours.)

Repeated heating of oils, as when frying, also causes oxidation of the fatty acids but the rancid flavour tends to boil off. Another effect caused by the heat is that the fatty acids tend to link together (polymerize) and darken in colour. At this stage the oil should be discarded.

Although polyunsaturated fatty acids would be expected to oxidize very easily because they have so many double bonds, it has been shown that one can use polyunsaturated oils like safflower, sunflower, maize, and soya for 5 to 10 batches of chipped potatoes without losing more than 10% or so of PUFA. So people who have been medically advised to eat polyunsaturated fats do not need to discard their frying oil each time it is used.

Margarine is made mostly from liquid vegetable oils and fish oils by a process of hardening. This is done by treating the oils with gaseous hydrogen in the presence of a metallic catalyst. In this way hydrogen is added to the double bonds and the fatty acids become saturated and hard. In order to make semi-soft margarines which are rich in polyunsaturates a special process has to be used in which untreated oils very rich in PUFA are blended with partially hardened fats.

ENERGY – CALORIES AND JOULES

The energy value of fuels is measured in units of heat called calories. A calorie is defined as the amount of heat needed to raise the temperature of a gram of water 1 °C.

Until recently the nutritionist and the engineer measured the energy content of foods and fuels and the energy output of the body and machines in the same heat units. Since the nutritionist deals with relatively large amounts it was more convenient to use the kilocalorie (kcal) or Calorie with a capital C, which is 1 000 (small) calories.

In 1969 the International Union of Nutritional Sciences recommended that the joule, which is a unit of energy, should replace the calorie, which is a unit of heat. Adoption of this was recommended in the UK by the Royal Society in 1972. For several years both units will continue to be used side by side.

The joule (pronounced jool) is derived from the (small) calorie by multiplying by 4·18 (to be precise 4·184).

Thus 4·18 J = 1 cal
4·18 kJ (kilojoules) = 1 kcal (note small k)
4·18 MJ (megajoules) = 1 000 kcal (note capital M)

An easy way to remember is to keep in mind the daily energy needs of the average adult: 2 500 kcal = approx. 10 MJ.

Note that terms such as calorie content of foods and calorie intake would be expressed as energy content and energy intake.

Carbohydrates vary slightly from one another in energy content simply because of the relative amounts of carbon, hydrogen and oxygen. For example glucose and fructose contain 3·75 kcal/g, sucrose 3·95 and starch 4·1. These differences are usually ignored, especially since most of the dietary carbohydrate is starch, and the energy content of carbohydrates is taken as 4 kcal/g.

However, the easiest way to measure the total carbohydrate in a food (starch plus sugars) is to hydrolyse all of it to glucose and fructose, and estimate these two. This is the meaning of the term used in Food Composition Tables 'carbohydrates expressed as monosaccharides'. Then the amount of monosaccharide is multiplied by 3.75 (not 4) to find the calorie content of the carbohydrates in the food.

Expressed as joules this would be 3·75 x 4·18 = 15·7 kJ/g; rounded off to 16.

Proteins provide 4 kcal or 16·7 kJ, rounded off to 17 kJ/g.

Fats average 9 kcal/g = 37·6 kJ. This figure of 9 kcal is believed to be higher than has been measured in human subjects (there is very little accurate information) so the Royal Society recommends that the value of 37·6 kJ should be approximated down to 37.

Energy Values	*per gram*		*per oz (rounded off)*	
	kcal	*kJ*	*kcal*	*MJ*
Carbohydrate (as monosaccharides)	3·75	16	110	0·45
Protein	4	17	115	0·48
Fats	9	37	260	1·05
Alcohol	7	29	200	1·20

ENERGY VALUE

The energy value of foods is calculated simply by finding the amounts of carbohydrate, fat and protein present, either by analysis or from food composition tables, and multiplying each by its appropriate factor. A food that contains a great deal of water will naturally be of low energy value while one rich in fat will be of high energy value (Fig. 7).

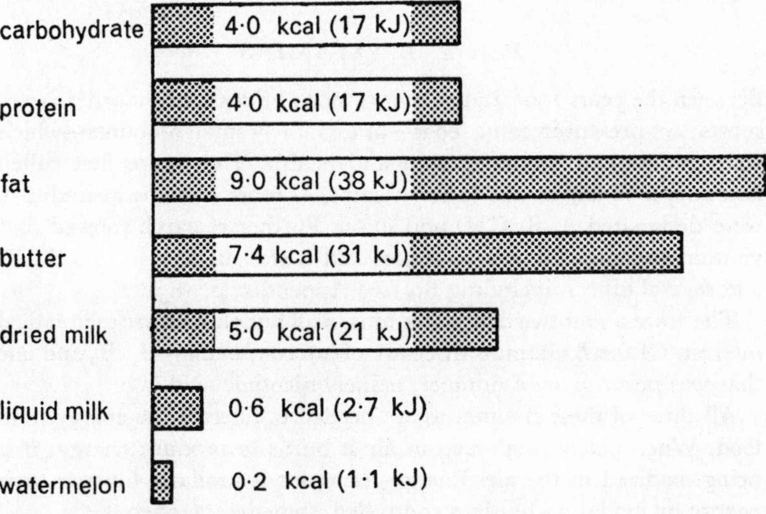

Fig. 7 Energy values (kcal and kJ per gramme)

When foods are fried they absorb part of the fat and thereby increase in energy value (Fig. 8). For example, raw potatoes supply 87 kcal per 100 g (25 kcal per oz) and when they are converted into chips by frying they supply 239 kcal per 100 g (70 kcal per oz). They have absorbed fat and lost some water. Some foods can soak up relatively large quantities of fat and change from low-energy to high-energy foods.

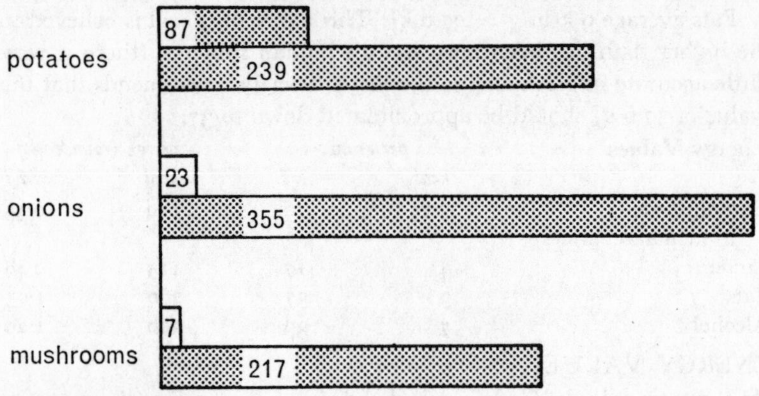

Fig. 8 Increase in energy value of food on frying (kcal per 100 g)

For example, raw onions provide 23 kcal per 100 g but after frying the figure becomes 355 kcal. Similarly mushrooms increase from only 7 to 217 kcal (from 29 to 908 kJ per 100 g).

Part 2 B VITAMINS

Between the years 1900 and 1910 it came to be realized that there were substances present in some foods—in extremely small amounts—which were essential to proper growth and health. They were first called accessory food factors and later vitamins. As more were discovered they were designated A, B, C, D and so on. Further research showed that vitamin B contained several factors and it was subdivided into B_1, B_2 and several others, including B_{12} (see Appendix, p. 183).

There are about two dozen vitamins in all but only six are of practical interest. Of the B vitamins three are of interest, namely B_1, B_2 and one that was never given a number, namely nicotinic acid.

All three of these B vitamins are needed to liberate the energy from food. When petrol is sparked in air it burns to produce energy; it is being oxidized in the air. Energy-giving foods similarly liberate their energy by oxidation but in a controlled, stepwise, manner.

VITAMIN B_1 OR THIAMIN (AT ONE TIME CALLED ANEURINE)

Thiamin assists the oxidation of glucose in the cells of the body. Normally, glucose is oxidized completely to carbon dioxide and water and its energy liberated. If thiamin is in short supply then the process is interrupted: the glucose is only partly metabolized and the breakdown stops at pyruvic acid formation. This accumulates in the blood. So far as the individual is concerned he suffers from muscular weakness, palpitations of the heart and degeneration of the nerves. The disease is called beri-beri and is still common in some parts of the Far East and other parts of the world where the diet is deficient in thiamin.

Since thiamin is needed to assist the oxidation of glucose the amount needed depends on the amount of carbohydrate in the diet. The recommended allowance is 0·6 milligrammes for every 1 000 kcal derived from carbohydrate or approximately 0·4 mg per 1 000 total kcal. So the average adult man engaged in light work and spending 2 500 kcal per day would need 1 mg of thiamin (see Table 1). A surplus is harmless since it is excreted in the urine.

To obtain enough thiamin in the daily diet an adult could eat 5–6 oz (150 g) of bread, 2 oz (57 g) bacon or oatmeal or ¾ oz wheatgerm, and 10 oz of potatoes. Small amounts would be supplied by milk and green vegetables.

Like all the B vitamins, thiamin can fairly readily be washed out of food into the cooking water, particularly if the food is finely chopped,

TABLE 1

Recommended intakes for B vitamins (F.A.O.)

	Thiamin (mg)	*Riboflavin (mg)*	*Niacin (mg)*
Children 0 to 1 year	0·4	0·55	6·6
1 to 3	0·5	0·7	8·6
4 to 6	0·7	0·9	11·2
7 to 9	0·8	1·2	13·9
10 to 12	1·0	1·4	16·5
Boys 13 to 15	1·2	1·7	20·4
16 to 19	1·4	2·0	23·8
Adult men	1·3	1·8	21·1
Girls 13 to 15	1·0	1·4	17·2
16 to 19	1·0	1·3	15·8
Adult women	0·9	1·3	15·2
pregnancy	1·1	1·6	18·5
lactation	1·3	1·9	21·8

Table 2
Good sources of vitamin B_1
Portions supplying ⅓ of the daily needs of the adult (about 0·3 mg)

Food (raw)	*Portion*	*Food*	*Portion*
Heart (ox)	65 g (2¼ oz)	Oatmeal	60 g (2 oz)
Bacon (lean)	45 g (1½ oz)	Flour, wholemeal	65 g (2¼ oz)
Pork (lean)	30 g (1 oz)	white (enriched)	90 g (3 oz)
Pork (fat)	70 g (2½ oz)	Soya flour	40 g (1½ oz)
Cod's roe	20 g (¾ oz)	Brazil nuts	30 g (1 oz)
Liver and kidney (average)	100 g (3½ oz)	Yeast, dry	15 g (½ oz)
		brewers'	3 g (1/10 oz)
Dried pulses	65 g (2¼ oz)	Bread, wholemeal	115 g (4 oz)
Peanuts	30 g (1 oz)	white (enriched)	200 g (7 oz)
Potatoes	300 g (10½ oz)	Wheatgerm	20 g (¾ oz)
		Bran	30 g (1 oz)

but, unlike the other B vitamins, it is damaged by heat and by alkalis. To reduce losses cook quickly. For example pulses and bacon should be soaked then steamed or pressure-cooked, pork should be tenderized by beating, treating with a tenderizer, marinading, or cutting against the grain as in the Chinese 'stir-fry' pork. If the soaking or cooking water can be used in other dishes, losses are further reduced.

Rice Much of the thiamin of rice occurs in the outer husk and germ. This is lost when brown rice is husked in the preparation of white or polished rice. The loss is of little significance in the Western mixed diet but is serious when rice forms the major part of the diet. If the rice is parboiled before removing the husk, part of the thiamin is rinsed into the grain and not lost with the husk. This is the method that should be used in areas where rice is the staple food.

VITAMIN B_2 OR RIBOFLAVIN

This vitamin also plays a part in the oxidation of glucose in the cell. The amount needed is, therefore, related to the amount of carbohydrate that is used and the recommended allowance for the average adult is 1·5 mg per day. The lips, tongue and skin are affected when there is a shortage of riboflavin in the diet—'ariboflavinosis'.

Good sources of riboflavin are milk and milk products—sheep's milk is nearly four times as rich as cow's milk—liver, kidney and yeast.

Milk The riboflavin of milk is slowly destroyed by light at the rate of about 10% every hour, and much of the riboflavin content of milk can be lost if it is left in bright sunlight for long periods. At the same time as the riboflavin is destroyed by light, the vitamin C is also destroyed but at a much faster rate. So, for the sake of these two vitamins, milk bottles should be kept in the shade or the milk protected by a carton or dark glass bottles.

TABLE 3

Good sources of vitamin B_2

Portions supplying $\frac{1}{3}$ of the daily needs of the adult (about 0·5 mg)

Food	*Portion*	*Food*	*Portion*
Milk, cow's	260 g ($\frac{1}{2}$ pt)	Skim milk powder	30 g (1 oz)
goat's	330 g (12 oz)	Egg	105 g ($3\frac{1}{2}$ oz)
Hard cheese	100 g ($3\frac{1}{2}$ oz)	Almonds	60 g (2 oz)
Liver, ox (raw)	15 g ($\frac{1}{2}$ oz)	Yeast, dried	15 g ($\frac{1}{2}$ oz)
Kidney, ox (raw)	30 g (1 oz)	brewers'	15 g ($\frac{1}{2}$ oz)
Tea, extract from	40 g ($1\frac{1}{2}$ oz) of leaves	Yeast extract	5 g ($\frac{1}{5}$ oz)
Meat extract	7 g ($\frac{1}{4}$ oz)	Soft curd cheese	260 g (9 oz)

NIACIN (NICOTINIC ACID AND NICOTINAMIDE)

The third member of the B vitamins is found in food in two forms, nicotinic acid or its amide, nicotinamide, both referred to as niacin. It can also be formed from the amino acid tryptophan; hence the term 'niacin equivalent' is often used. This is the niacin in the food plus one sixtieth of the tryptophan, since 60 mg of the amino acid can form 1 mg of the vitamin.

Niacin, like thiamin and riboflavin, plays a part in the release of energy from foods. A shortage leads to muscular weakness, mental and digestive disorders and dermatitis, and is called pellagra.

Pellagra occurs in most maize-eating countries for two reasons. Firstly the vitamin is chemically bound and not available to the body; secondly, maize is low in tryptophan content.

An example of how traditional cooking practices can improve nutritional values is the treatment given to maize in Mexico. Before making tortillas, a kind of pancake, the maize is soaked overnight in lime water. This liberates the bound niacin and makes it available—and pellagra is rare in Mexico.

Since it is involved in carbohydrate metabolism the requirements of niacin are related to the amount of carbohydrate consumed. For the average adult engaged in light work the recommended allowance is 18 mg per day. Foods rich in this vitamin are meat, whole-grain cereals and enriched milled cereals, nuts, liver, yeast and yeast and meat extracts.

Niacin is stable to heat, alkali, and light; but like the other B vitamins it can be leached out of chopped foods into the cooking water.

TABLE 4
Good sources of niacin
Portions supplying $\frac{1}{3}$ of the daily needs of the adult (6 mg)

Food	*Portion*	*Food*	*Portion*
Peanuts and peanut butter	30 g (1 oz)	Fungi, dried	85 g (3 oz)
Wholemeal bread	105 g ($3\frac{1}{2}$ oz)	Liver, ox	30 g (1 oz)
White bread (enriched)	200 g (7 oz)	Barley, pearled	125 g ($4\frac{1}{2}$ oz)
Yeast, dried	15 g ($\frac{1}{2}$ oz)	Coffee	20 g ($\frac{3}{4}$ oz)
brewers'	15 g ($\frac{1}{2}$ oz)	instant	15 g ($\frac{1}{2}$ oz)
Rice, white	200 g (7 oz)	Beer, mild	1 litre ($1\frac{3}{4}$ pt)
boiled	750 g (26 oz)	Pulses	100 g ($3\frac{1}{2}$ oz)
Soya flour	60 g (2 oz)	Sunflower seed	100 g ($3\frac{1}{2}$ oz)
Meats, average (raw)	140 g (5 oz)	Yeast extract	10 g ($\frac{1}{3}$ oz)
		Meat extract	7 g ($\frac{1}{4}$ oz)
		Bran	20 g ($\frac{3}{4}$ oz)

DO WE GET ENOUGH OF THE B VITAMINS?

In Great Britain, except in very rare and exceptional circumstances, there are no signs of diseases such as beri-beri and pellagra. Dietary surveys show that the *average* intake of the B vitamins is greater, although not markedly greater, than the recommended allowances. There are people living on restricted diets who might be on the borderline; and we must bear in mind the vulnerable groups who have enhanced requirements for nutrients, such as children, adolescents, and pregnant women and nursing mothers.

Individuals with a 'sweet tooth' may be consuming relatively large amounts of carbohydrate as sugar without the B vitamins needed to metabolize the sugar. It seems necessary, therefore, to ensure that all diets are planned to supply an adequate amount of the B vitamins and to keep storage and cooking losses to a minimum.

WHAT CAN BE DONE?

Some foods, such as wholemeal cereals, peanuts, soya, wheatgerm, yeast, pork and bacon, are exceptionally good sources of all three of the

B vitamins. During the milling of wheat to produce the ordinary white flour of 72% extraction rate, some of the B vitamins are lost with the bran and germ. In Great Britain two of them, thiamin and nicotinic* acid are added back to all white flour that is sold, and they are, therefore, also present in all white bread. The amount added back to this enriched flour makes it equivalent in these vitamins to 80% extraction flour. In some countries riboflavin is also added.

Some foods, such as heart, kidney and liver, are so rich a source of these vitamins that a regular, say weekly, dish would go a long way towards ensuring an adequate intake. Dishes can be enriched with dried skim milk powder, soya and brewers' yeast.

As stated earlier the major source of man's food is cereal or starchy root crops. Many of these foods contain enough of the B vitamins to metabolize the glucose they provide. However, as we prepare refined foods, such as polished rice and white flour some of the B vitamins are lost and there may no longer be sufficient in the particular food to metabolize the glucose. This may not be important where there are many other foods available to provide the missing vitamins; and it is always possible, as is done in many countries, to restore the missing vitamins. However, the vitamin loss becomes of vital importance where the food in question is providing as much as two-thirds or even three-quarters of the total diet and there are very few vitamin-rich foods available to make up for the deficiency. It is also of importance when, as has happened in so many countries, the consumption of sugar has increased so much—it has already become a fifth of the average diet in Great Britain—because sugar (and there is no real difference between white and brown) provides calories only and no vitamins at all—the so-called empty calories.

Some traditional dishes based on white rice include a rich source of the vitamins. For example, the Chinese add pork, liver and nuts, or dried fungi, and they make use of soy sauce and soy bean curd. Kitcheree, which is white rice cooked with lentils, is both savoury and adequate in B vitamins. Brown rice is not always popular but might be acceptable in a pilaff with meat, coloured brown with yeast or meat extract. Home-made bread and beer are useful sources of B-vitamins.

GETTING ENOUGH

The daily needs for B vitamins could be supplied by: 10 oz (280 g) bread, and 2 oz (57 g) meat, 2 oz (57 g) bacon *or* 3 oz (85 g) oatmeal

* The Food Standards Committee (1974) recommended that enrichment of flour with niacin be discontinued.

and 1½ oz (42 g) of peanuts, plus ⅔ pt (0·38 l) of milk or ¾ oz (20 g) liver (see also Fig. 9) or 1½ oz bran.

A weekly intake of 8 oz (230 g) of liver or kidney or 3½ oz (100 g) of wheatgerm would go a long way towards ensuring an adequate intake of the B vitamins.

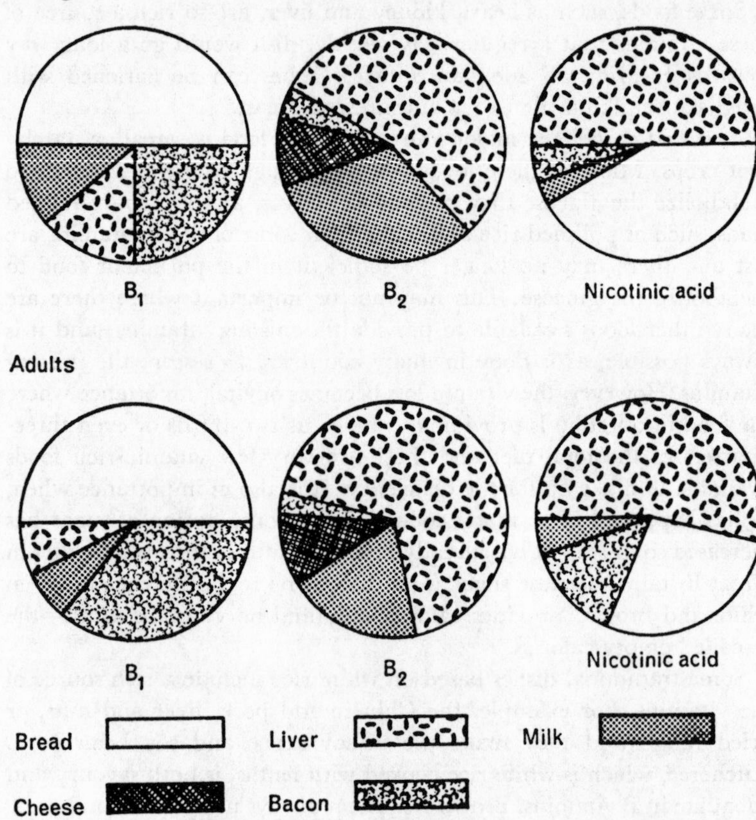

Fig. 9 Proportions of daily requirements of B vitamins derived from foods suggested on page 17

Since B vitamins are needed to metabolize carbohydrate, the greater the energy intake from carbohydrate, the higher are the needs for the B vitamins. However, if the foods themselves are adequate in vitamin content, i.e. contain the right amount of B vitamins per 1 000 kcal (0·4 mg B_1, 0·6 mg B_2, 7·0 mg niacin) then this is automatically taken care of when more food is eaten.

RECIPES (starred dishes, see page 193)

1.* Stir-fry pork.
2. Cheese bread (with brewers' yeast and cheese).
3. Peanut bread.
4. Ground nut stew (meat and vegetable stew thickened with peanut butter).
5.* Taramasalata (*hors-d'œuvre* using smoked cod's roe).
6. Chow nap nap (pork, nuts, sweet corn).
7.* Wholemeal bread and rolls.
8.* Muesli or summer breakfast (oats, nuts, milk, apples).
9. Home-made sausages.
10. Danish bacon roast.
11. Cornell bread (enriched with dried milk, yeast, wheatgerm, egg).
12. Kitcheree (rice cooked with lentils).
13. Pork and beans (soaked) pressure-cooked with tomatoes, for 20 min.
14.* Sweet-sour pork.

Part 3 ENERGY NEEDS

HOW MUCH FUEL DO WE NEED?

Fuel is needed by our bodies for our everyday activities. But even when we are resting, or even asleep, our bodies are still using fuel to keep the heart beating, to maintain muscle tone and to keep us warm. So it is possible to divide fuel needs into two parts: that required when we are at complete rest, called the basal metabolism, and that required for activity, called work energy. Basal metabolism is defined as the energy or heat output when at complete rest (but not asleep), at least 12 hours after a meal (so there is no energy being used in digestion and there is no specific dynamic action (see later)) and in a warm room. This is called the basal metabolic rate or BMR.

Continuing the earlier analogy of the car, the amount of fuel depends on the horse power (BMR depends on body size) and the distance travelled (work energy depends on the amount of work done). If the car takes on more fuel than it is using the boot will soon be bulging with stored tins of surplus fuel. So with people: if they take in more fuel, as food, than they are using, they store the surplus as fat and they, too, begin to bulge.

ENERGY

We speak of the energy content of food, the energy needed to do work and the energy needed to produce heat. All three are basically the same and can be measured in calories or joules.

The amount of heat given out by the body, that is the energy expenditure, can be measured directly. The subject is confined in a small, completely insulated apparatus so that all the heat produced can be measured. However, as this is such a laborious process the heat output is usually measured indirectly.

When fuel, for example glucose, is burned oxygen is used and carbon dioxide is formed:

glucose + oxygen → carbon dioxide + water + energy

The amount of energy produced can be calculated either from the oxygen used or the carbon dioxide produced. All the subject has to do is to breathe into equipment which will collect and measure the gases.

BASAL METABOLIC RATE

BMR, the resting metabolism, depends mostly on the surface area of the body. This can be measured or may be calculated from height and weight according to a given formula. The amount of heat produced varies with age and sex and, if the surface area is known, the BMR can be calculated (see Fig. 10 and Table 5).

For example, suppose we wish to know the basal metabolic output of a 16-year-old girl, height 5 ft 1 in (155 cm) and weight 100 lb (45 kg). If a ruler is laid on Fig. 10 from 5 ft 1 in on the left line to 100 lb on the

TABLE 5
Basal metabolic rates
(energy output per m² body surface per hour)

Age	*Males*		*Females*	
	kcal	kJ	kcal	kJ
3	51·5–68·8	215–260	46·4–62·6	190–260
5	47·7–65·0	200–270	44·9–61·1	190–250
7	43·4–60·7	180–250	42·1–58·3	180–240
9	39·6–56·3	170–240	38·2–54·5	160–230
11	37·7–52·2	160–220	35·1–50·6	150–210
13	36·0–49·3	150–210	33·2–45·1	140–190
16	33·7–46·9	140–200	31·3–40·8	130–170
19	32·7–45·0	140–190	29·7–39·3	120–160
22	32·4–43·1	130–180	29·2–38·8	120–160
30	31·4–41·4	130–170	29·2–38·9	120–160
40	30·5–40·5	130–170	27·8–37·5	120–160
50	28·8–38·8	120–160	27·1–36·7	110–150
60	28·1–38·1	120–160	26·5–36·1	110–150
70	27·4–37·4	115–160	25·9–35·5	110–150

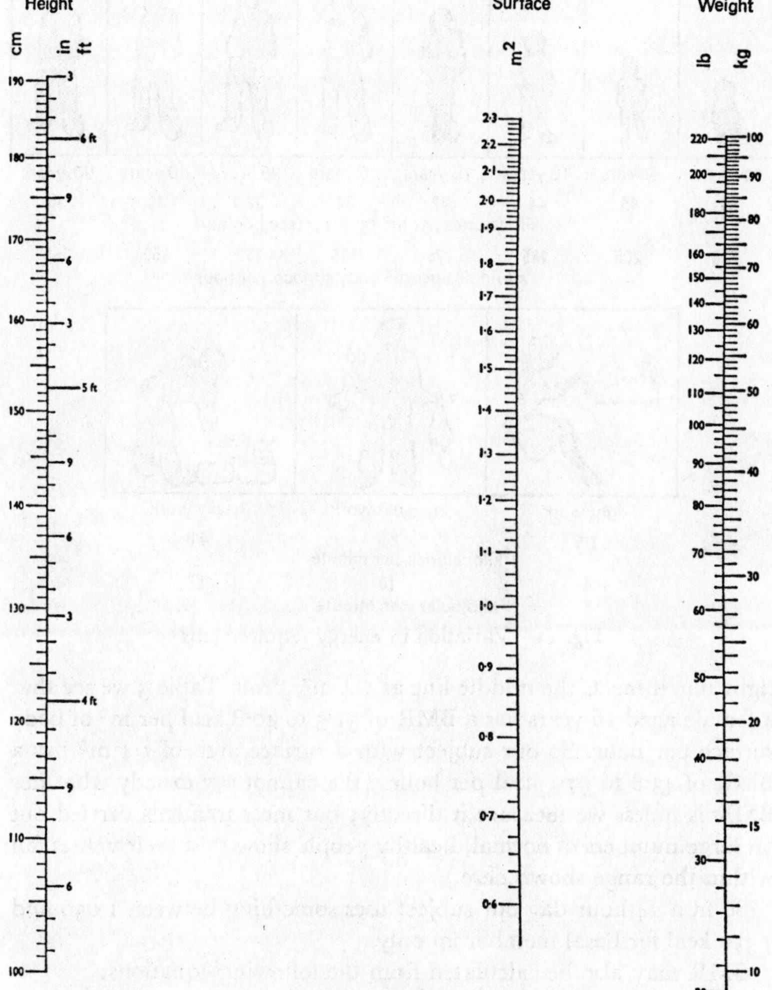

Fig. 10 Nomogram for the estimation of the surface area of the body from height and weight

Lay a ruler on the height of the subject (left line) and the weight (right line). The surface area in square metres can be read off where the ruler cuts the middle line.

The basal metabolic rate per square metre of body area is given in Table 5.

(The nomogram is constructed from the formula of Du Bois—Area = height to the power of 0·725 multiplied by weight to the power of 0·425 plus a factor of 71·84.)

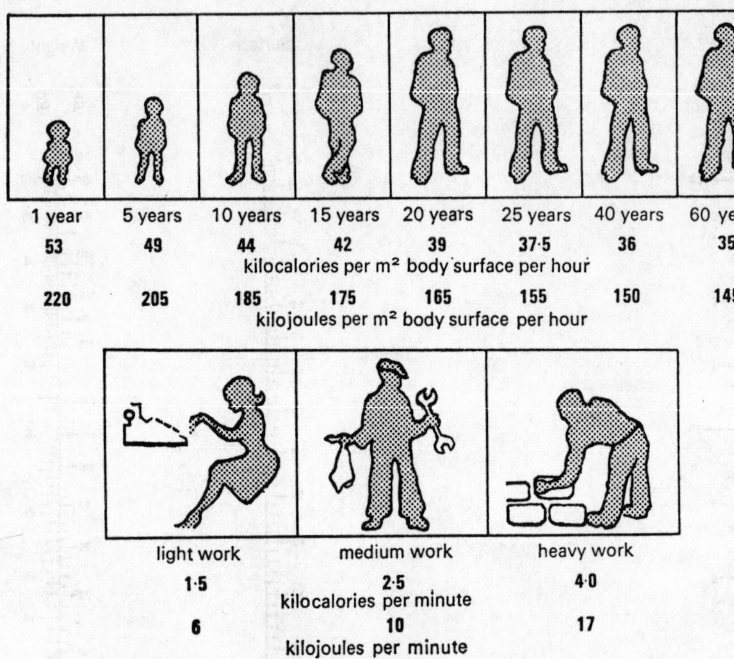

Fig. 11 Variation in energy requirements

right line it meets the middle line at 1·4 m^2. From Table 5 we see that a female aged 16 years has a BMR of 31·3 to 40·8 kcal per m^2 of body surface per hour. So our subject with a surface area of 1·4 m^2 has a BMR of 43·8 to 57·1 kcal per hour. (We cannot say exactly what her BMR is unless we measure it directly; but measurements carried out on large numbers of normal, healthy people show that their values fall within the range shown here.)

So in a 24-hour day our subject uses something between 1 050 and 1 370 kcal for basal metabolism only.

BMR may also be calculated from the following equations:

For men $E\text{ (kcal)} = 66{\cdot}5 + \{13{\cdot}7 \times W\text{ (kg)}\} + \{5{\cdot}0 \times H\text{ (cm)}\}$

For women

$$E = 66{\cdot}5 + \{9{\cdot}6 \times W\text{ (kg)}\} + \{1{\cdot}8 \times H\text{ (cm)}\} - \{4{\cdot}7 \times \text{age (yrs)}\}$$

As we become older there is a fall of about 5% in our BMR for every decade; so at 75 years of age our fuel needs to satisfy BMR may be only two-thirds of those at the age of 25—Fig. 12 and Table 6 (see also Chapter 6). In addition to this fall in BMR there will also be a fall in work energy as elderly people tend to work both less intensively and more economically.

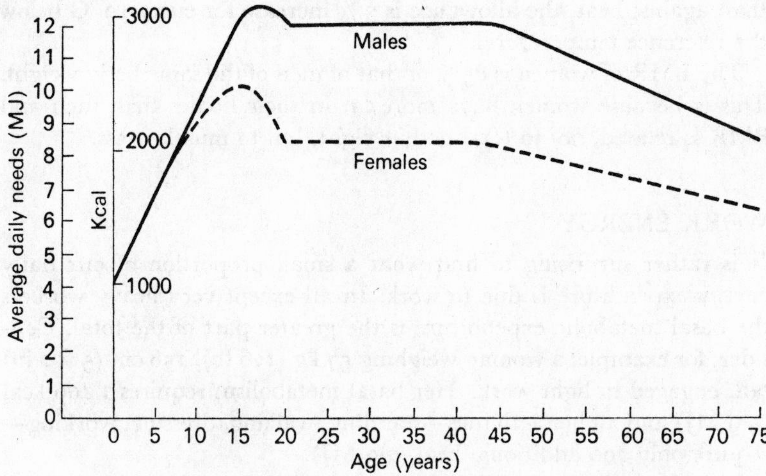

Fig 12 Change in energy requirements with age

TABLE 6
Energy requirements (Food & Agriculture Organization 1973)

	Males					*Females*				
Age	*Weight*	*Energy/ kg*		*Energy/ person*		*Weight*	*Energy/ kg*		*Energy/ person*	
		kcal	kJ	kcal	MJ		kcal	kJ	kcal	MJ
0–1	7·3	112	470	820	3·4	same				
1–3	13·6	100	418	1 360	5·7	same				
4–6	20·7	91	381	1 870	7·8	20·0	90	376	1 790	7·5
7–9	28·6	79	330	2 260	9·5	27·6	7·6	318	2 110	8·8
10–12	36·7	71	297	2 600	10·9	37·7	62	259	2 350	9·8
13–15	51·7	56	234	2 900	12·1	50·3	50	209	2 500	10·5
16–18	62·4	50	209	3 100	13·0	54·2	43	180	2 340	9·8
Adult										
20–39	65	46	192	3 000	12·6	55	40	167	2 200	9·2
40–49				2 850	11·9				2 090	8·7
50–59				2 700	11·3				1 980	8·3
60–69				2 400	10·0				1 760	7·4
70–79				2 100	8·8				1 540	6·4

As part of the heat production from our bodies is designed to keep us warm, climate affects heat losses. Calculations are usually made in relation to an outside air temperature of 10 °C. For every 10 °C above this reference temperature the heat output is reduced by 5%. At temperatures below the reference temperature there is an increased requirement. As we are more able to protect ourselves against cold than against heat, the allowance is 3% increase for every 10 °C below the reference temperature.

The BMR of women is 85% of that of men of the same body weight. This is because women have more fat in their bodies than men and BMR is related, not to total body weight, but to muscle mass.

WORK ENERGY

It is rather surprising to find what a small proportion of our daily energy expenditure is due to work. In all except very heavy workers the basal metabolic expenditure is the greater part of the total. Consider, for example, a woman weighing 57 kg (126 lb), 158 cm (5 ft 2 in) tall, engaged in light work. Her basal metabolism requires 1 400 kcal (5·9 MJ) and all her activities—standing, walking, dressing, working—require only 700 additional kcal (2·9 MJ).

This figure becomes of great interest when we discuss methods of losing surplus body weight, which means using up some of the stored fat. If the woman described above increased her work each day by 10% this would represent 70 extra kcal—which is only as much as is present in a single slice of dry bread.

The energy spent on active work depends on the intensity of the physical effort and the length of time. Table 7 shows the rate of energy expenditure per minute in different types of activity and how this is multiplied by the time of the effort in order to arrive at the total energy expenditure.

These figures are for total energy for each type of work, i.e. they include basal metabolism.

The cost per minute for some other types of activity is: light industrial work, most domestic work, carpentry, brick laying—2.5–4.9 kcal (10–20 kJ); gardening, tennis, dancing, cycling up to 20 km/h–5.0–7.4 kcal (21–30 kJ); coal mining, squash, cross-country running—more than 7·5 kcal (30 kJ).

INDIVIDUAL VARIATION

For the purpose of assessing energy needs of communities the standard reference man is taken as being 20–39 years of age, weighing 65 kg, engaged for 8 hours a day in moderate activity and living at a mean average temperature of 10 °C. The standard reference woman is living under similar conditions but weighs 55 kg.

However, even if you correspond exactly to the reference individual your energy expenditure and energy needs may be very different. This is simply the well-known phenomenon of biological variation. Some people are healthy and active on only half the food that other people need. So all tables of food allowances apply to 'average' individuals and to groups of the population, but not to single individuals. They are of great value in calculating the food needed in a school, factory canteen or in planning for a whole community; but they are not applicable to the individual.

TABLE 7

Daily energy balance sheet of a clerk, body weight 66 kg, height 168 cm

	Length of time (rounded off)	*kcal/min*	*Total per day (mean of 7 days' measurements) kcal*	MJ
Sleep	8 h	1·1	530	2·2
Off Work				
Light sedentary activity	4 h	1·48	360	1·5
Washing and dressing	½ h	3·0	90	0·4
Light domestic work	1 h	3·0	180	0·8
Walking	1 h	6·6	400	1·7
Gardening	25 min	4·8	120	0·5
Standing	20 min	1·56	30	0·1
At Work				
Sitting activities	3 h 10 min	1·65	320	1·3
Standing activities	3 h 40 min	1·9	420	1·8
Walking	12 min	6·6	80	0·3
	Sleeping		530	2·2
	Total working		820	3·4
	Total off work		1 180	4·9
	Total		2 530	10·6
		(BMR	1 680)	(7·0)
	Energy intake calculated from diet		2 620	11·0

Daily energy balance sheet of a coalminer

	kcal	MJ
Sleeping	530 kcal	2·2 MJ
Work	1 700 kcal	7·1 MJ
Remainder	1 560 kcal	6·5 MJ
Total	3 790 kcal	15·8 MJ
(BMR	1 680 kcal)	(7·0 MJ)
Energy intake calculated from diet	3 990 kcal	16·7 MJ

SPECIFIC DYNAMIC ACTION

If you calculate that an individual needs, say, 2 000 kcal (8·4 MJ) you will find that you must give him about 5 to 7% more than this. This is because the effect of supplying nutrients to the body is to stimulate metabolism and cause a 5 to 7% wastage as heat. This is called specific dynamic action or SDA.

FOOD COMPOSITION TABLES

Although proteins are required for body construction, they can also be used, together with fats and carbohydrates, as a source of energy. The energy content of a food can be measured directly by burning it in a bomb calorimeter and measuring the heat produced (then making allowances for the undigested portion) or by calculation from the amounts of fat, protein and carbohydrate present. These amounts can be found by analysis or in food composition tables.

Most foods have been analysed and their energy and nutrient contents listed in Food Composition Tables per 100 grammes and per oz. It is necessary only to know how much food is eaten in the dish or the meal or per day to calculate the intake of energy and the various nutrients.

As described on pp. 15 and 16 carbohydrates provide 3·75 kcal or 16 kJ per g; proteins provide 4 kcal or 17 kJ per g; and fats provide 9 kcal or 37 kJ per g.

Tables usually give values in terms of the edible portion of the food, i.e. any waste has already been rejected. If the calculation is made on, say, potatoes 'as purchased', i.e. with their peel, then the 25% wastage is already taken care of in the tables. They show that potatoes 'as purchased' provide 16 kcal per oz, so 100 kcal portion is 100/16 = approx. 6 oz.

If the calculation is to be made on 'edible portion', i.e. after peeling, this is shown in the tables as providing 21 kcal per oz, so 100 kcal portion in this case is 100/21 = approx. 5 oz.

Food tables are compiled from average results of analyses of a number of samples of the food in question. Foods can vary a lot between different samples in the case of, say meat or fish, and can vary a little between crops grown in different areas under different conditions, so food tables can provide only an approximation to the composition of the particular sample under consideration. The only alternative is a lengthy analysis. However, many foods such as milk, butter, sugar, bread, cereals, are of fairly constant composition, so that the error introduced by using tables instead of analysis is, overall, between 5 and 10%.

It is useful to be able to make a quick, even if approximate, assessment of the energy content of a dish or a meal. An excellent method of learning how to do this is to prepare portions of foods that provide 100 kcal or 0·4 MJ and note how the portions vary. A small pat of butter (½ oz or 14 g), a thick slice of bread (1½ oz or 42 g) and a large piece of cabbage (1 lb or 450 g) all provide the same amount of energy.

THE SIZE OF A MEAL

Some foods are not essential for the functioning of the body but we do enjoy eating them. Such foods, which could be called fun foods, can make a major contribution to the energy intake without making any contribution to the intake of other nutrients.

In a day we might drink 6 or 7 cups of tea, each with 2 lumps of sugar; this would amount to 200 kcal (and many people use twice as much sugar); eight sweet biscuits (400 kcal), 2 oz (57 g) of chocolate (310 kcal), a tube of peppermints (1½ oz = 150 kcal) and two glasses of lemonade (200 kcal)—all this adds up to 1 200 kcal, which is more than half the total day's energy requirements of the average woman. She will have to find most of her protein, mineral, and vitamin supplies in the remaining half of her food.

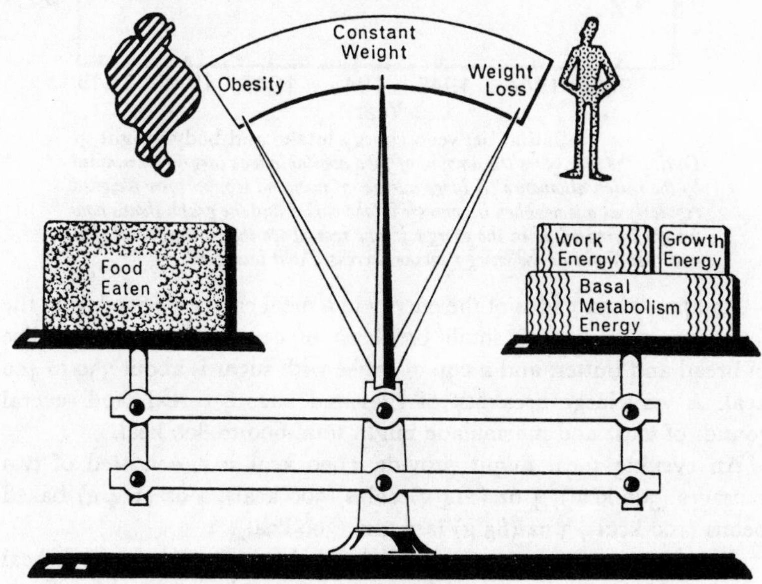

Fig. 13 Energy balance

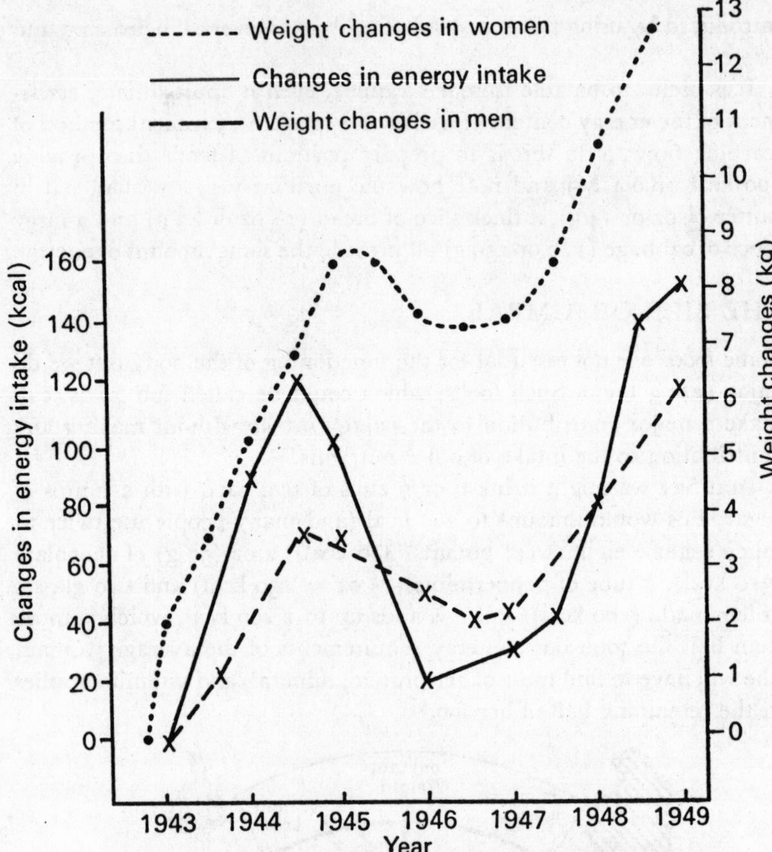

Fig. 14 Relation between energy intake and body weight
During the war years the amount of food available was largely determined by the ration allowance. A large number of men and women were weighed regularly at a time when the energy intake varied and the graph shows how weight increased when the energy intake rose above the starting level, fell and rose again as the energy intake decreased and increased.

An approximate idea of the energy in a meal can be derived from the following examples. A small breakfast of cereals and milk, a slice of bread and butter, and a cup of coffee with sugar is about 350 to 400 kcal. A very large breakfast of eggs and bacon, cereal, and several rounds of toast and marmalade might total 600 to 800 kcal.

An evening meal might provide 1 000 kcal if it consisted of two sausages (300 kcal), 4 oz (114 g) chips (300 kcal), 4 oz (114 g) baked beans (100 kcal), 3 oz (85 g) jam tart (300 kcal).

An average woman engaged in light work requires about 2 100 kcal per day. If we allow 200 kcal for cups of tea and occasional fun foods and assume that breakfast is the smallest meal and the evening meal is the largest, this would mean about 400 kcal for breakfast, 600 for lunch and 800 in the evening. Certainly individuals vary; but if we look at the 800 kcal breakfast given earlier, and allow a lunch of the same size and a larger evening meal, we can see how easy it is to eat too much.

ENERGY BALANCE

When people remain fairly constant in body weight they must be expending as much energy as they are consuming, i.e. they are in energy balance (Figs. 13 and 14). The appetite usually ensures that we do satisfy our energy needs but it can be a very poor guide to ensuring that we do not overfulfil these needs. Energy intake surplus to requirements is stored in the body as fat. While this fat reserve is of great use to hibernating animals and to camels crossing the desert without food, human beings are well advised to adjust their intake to their needs because excessive weight throws a strain on the body and leads to ill-health. (See Appendix, Table 37, weight tables.)

PRACTICAL WORK

1. Weigh out and arrange an exhibit of 100 kcal portions of food. (0·4 MJ)
2. Group together foods useful in a low-energy diet.
3. Compare the energy content of savoury and sweet foods, e.g. two chocolates and an egg salad; two sweet biscuits and 10 oz grilled plaice; a piece of cake and poached egg with smoked haddock fillet.
4. Calculate the day's energy intake of an 18-year-old boy:
   a. 2 ham sandwiches (2 slices ham, 4 slices bread, 1 oz (28 g) butter and 1 pint (570 ml) beer.
   b. 2 rolls with butter and jam, 2 rock cakes, 1 chocolate biscuit, ½ pint (285 ml) milk.
   c. Glass of tomato juice, 1 oz (28 g) cheese, 2 plain biscuits.
   d. 2 rashers bacon, 1 egg, 2 slices bread, 7 lumps sugar.

   How adequate is this?
5. Calculate the energy per portion of these recipes and express them as a fraction of the daily needs of a woman at 2 100 kcal (9 MJ) per day.
   a. Victoria sponge—6 oz (170 g) flour, 4 oz (114 g) each butter and sugar, 2 oz (57 g) icing sugar, 2 eggs, 1 oz (28 g) walnuts to decorate—8 portions.
   b. Jam roll—8 oz (230 g) flour, 3 oz (85 g) suet, 6 oz (170 g) jam—4 portions.

c. Cheese sauce—2 oz (57 g) flour, 2 oz (57 g) butter, 1 pint (570 ml) milk, 6 oz (170 g) cheese—6 portions.

6. Plan two tray lunches of 500 kcal (2·1 MJ) value each suitable for café service for office staff.
7. How would you increase the energy value by about 300 kcal (1·2 MJ) of (a) bowl of soup; (b) plate of mixed salad; (c) baked fillet of haddock?

EXAMINATION QUESTIONS

1. From what foods do we obtain energy? How is this energy used by the body? Explain the differences in diet for a convalescent and for an active teenager, giving examples of appropriate meals for each. (City and Guilds Adv. Dom. Cookery.)
2. Explain the body's needs for energy. Give typical energy requirements for three different groups of people. What are the causes of changes in the energy needs of individuals? (R.S.H.)
3. What kind of diet is likely to be low in the B-group of vitamins, and why? (R.S.H.)
4. What is the value of fat in the diet? Name two animal and two vegetable fats and state how they can be used in cooking. Compare the price and food value of butter and margarine. ('O' level, London.)
5. A group of young men are going on a day's expedition to climb a mountain. Assuming that they require about 4000 kcal each for the day, plan their menu, including at least one packed meal. (R.S.H.)
6. Discuss the significance of the term 'unsaturated' when applied to edible fats. (Isleworth Diploma in Home Economics.)
7. What are the food constituents chiefly responsible for the production of heat and muscular energy, and the origins of these constituents? What is a calorie? Explain briefly the difference in the calorie requirements of individuals. (City and Guilds Dom. Cookery.)
8. How would you decide (a) if a food is a good source of thiamin? and (b) if a diet is adequate for this vitamin? Give three recipes improving the thiamin value of instant potato, white rice, sugar.
9. Describe the problems you would face in using Tables of food composition to calculate the nutrient content of a meal consisting of:

   Steak and kidney pie, boiled potatoes, boiled Brussels sprouts, stewed apricots and custard. (R.S.H.)

3 Foods for vitality

Having vitality means being lively as well as energetic. To produce energy to live, muscle cells, brain cells, and *all* body tissues need oxygen as well as food materials. Life is based on the oxidation of foodstuffs. So there must be an efficient system to supply enough oxygen to all body tissues.

It is the blood that carries oxygen to the cells and tissues, so any inadequacies in the blood system will have a serious effect on the functioning and vitality of the body.

Part 1 IRON

Blood consists of red and white cells suspended in a clear, yellow liquid, the plasma. The red cells (erythrocytes) are the means of carrying oxygen from the lungs to the tissues, and carbon dioxide (which is a waste product formed as the result of the oxidation of foodstuffs) from the tissues to the lungs where it is breathed out. The white cells (leucocytes) are concerned with the destruction of any bacteria that enter the body. The plasma is a solution of proteins (albumin and globulins) and other materials and is also the means of carrying the blood sugar and fats to and from the tissues.

The red cells contain a red pigment called haemoglobin (from the Greek, *haima*, blood). It is the haemoglobin that carries the oxygen. When blood goes from the heart to the lungs the haemoglobin picks up oxygen. From the lungs the blood returns to the heart and is then pumped through the arteries to all the organs and tissues. Here the oxygen is removed and in its place, the waste carbon dioxide is picked up. The blood carrying the carbon dioxide travels back to the heart in the venous system, and is then pumped to the lungs where the carbon dioxide is breathed out and replaced by fresh oxygen. Hence an adequate amount of haemoglobin is essential to the proper functioning of the body.

BLOOD FORMATION

Red blood cells are made in the bone marrow and a number of

nutrients are needed. The haemoglobin itself contains iron, so both iron and protein are needed. In addition, a number of other nutrients are required in very small amounts: the minerals copper and manganese are needed in such small amounts that there is no need to pay any attention to them in the diet, everyone has enough; vitamin C is needed to help the absorption of the iron from the food into the bloodstream; and there are three other vitamins which are usually plentifully supplied in most diets, vitamin B_{12}, folic acid, and vitamin B_6.

If there is a shortage of any of the factors needed for blood formation then anaemia, which simply means shortage of blood, results. There are many different kinds of anaemia but the commonest one is iron-deficiency anaemia, also called nutritional anaemia. (Pernicious anaemia has quite a different cause: it is a result of a failure to absorb the vitamin B_{12} from food, so the remedy is to give the vitamin by injection.) Since B_{12} is found only in animal foods, people living on wholly vegetable diets, e.g. Vegans, are in danger of anaemia from shortage of this vitamin.

ANAEMIA

People who suffer from anaemia, whatever the cause, cannot supply enough oxygen to their tissues. Although the body compensates to some extent by increasing the blood flow, these people are lethargic, sluggish, and easily fatigued. They find it difficult to concentrate, they get out of breath on even slight exertion, and they lack 'vitality'. The shortage of the red haemoglobin makes them look pale.

Iron-deficiency anaemia is all too common even in this country where the diet, in general, is very good. About one-sixth of the women in Great Britain suffer from anaemia, largely owing to the monthly blood loss which results in a loss of iron from the body. Therefore, they have a greater need for dietary iron than men do, and anaemia is found much more rarely in men.

The problem is that only a small part, about 5 to 20%, of the iron in our diet is absorbed from the food into the body. On the other hand the body normally conserves its iron very well. A red blood cell lasts 120 days and then it is broken down, but the iron is not lost, it is kept in the body for making new red cells. So normally the body needs only a milligramme of iron per day. Anaemia occurs mainly when there is a prolonged loss of blood, or failure of absorption or a dietary shortage over a very long period.

The amount of haemoglobin in the blood of healthy men is between 14 and 18 grammes per 100 ml of blood, with an average of 15·8 g. For women the range is 12 to 15·5 g with an average of 13·7 g. For convenience the figure of 14·8 g is taken as the standard and haemoglobin levels are calculated as a percentage of this figure of 14·8. Thus, a person with 7·4 g of haemoglobin per 100 ml of blood is said to have a haemoglobin level of 50%. This would be severe anaemia.

Obviously, it is not easy to say exactly when a person has anaemia. To be a little below the so-called 100% level is still within the normal range of many healthy people; some, from the above figures, can have a haemoglobin of well over 100%. Where, then, can a line be drawn? The figure suggested by the World Health Organization is 13 g for men, 12 g for women. It is on this basis that one-sixth of the women in Great Britain are said to be anaemic. The anaemia develops slowly over a period of years and anaemic people often say they feel 'perfectly well' because they have forgotten what it feels like not to be anaemic.

TREATMENT AND PREVENTION

As only a small part of the iron in our food is absorbed into the blood, the dietary treatment of anaemia may be a long slow process. Clearly, iron-rich foods should be eaten but the process is speeded up if iron-containing tablets, such as ferrous sulphate, are taken as well. But even with this medication it may take months to restore the iron loss. When the anaemia is very severe the remedy may have to be injections of iron, as in this way large doses can get into the body by by-passing the digestive system.

Prevention is better and easier than cure. If we eat foods rich in iron regularly then anaemia will not develop (unless there is some disease or persistent haemorrhage).

The U.K. recommended daily intakes are as follows:

Age	*Males*	*Females*
3–7	8 mg	8 mg
9–12	13	13
15–18	15	15
Adult	10	12
Pregnancy and lactation		15

These are the amounts that should be taken in the diet; that is to say the fact that only about one-tenth is absorbed has already been allowed for. The amount that is absorbed seems to vary considerably with different individuals and in the same individual at different times depending on other ingredients in the diet. Substances that make the iron insoluble in the intestine prevent its absorption: for example, phosphate, oxalate (from rhubarb and spinach), and phytate (from cereal fibre) can all combine with iron to make it insoluble.

Some of the iron present in food is not available to the body until it has been acted on by the hydrochloric acid of the stomach, so that people with low stomach acidity will absorb less of the iron. On the other hand, the presence of vitamin C enhances the absorption. People short of iron absorb more than healthy people (although obviously they still do not absorb enough to cure the anaemia).

This great variability in the amount of iron absorbed makes it very difficult to know whether people are getting enough.

WHICH ARE THE BEST FOODS FOR IRON?

Rich sources may be regarded as those containing more than 2 mg of iron per oz of food or 7 mg per 100 g, such as cocoa, liver and kidney (4 mg per oz), dried fish (2·5 mg), treacle (2·6 mg), shellfish such as mussels and cockles (3 to 7 mg), sesame seed, soya flour and pulses (average 2 mg per oz dry weight). Bread is fortified with iron in many countries as are several manufactured foods. Green vegetables are a moderate source.

Poor sources of iron are milk, cheese, sugar, sweets, jams, unfortified milled cereals such as wheat flour and maize starch (corn-flour), arrowroot and rice. It should be noted that gastric diets made from easily digested bland foods such as white fish and milk are very low in iron (and in vitamin C content).

Table 8 can be used to plan meals and diets.

PHYTIC ACID

One of the substances which makes iron insoluble and consequently poorly absorbed is phytic acid. This is found in wholemeal cereals, wheatgerm, bran and pulses. Fortunately, methods of food preparation can help. Phytic acid is broken down by yeast and so wholemeal bread made with a long rise has more iron available than, for example, soda bread and scones. White flour has most of the phytic acid removed in milling but is also low in iron content unless it has been fortified. The iron of oatmeal porridge becomes more available if it is soaked overnight (this shortens the cooking time) and eaten with milk, as any phytic acid left combines with the calcium of the milk and spares the iron. Pulses are improved in this respect by soaking in soft water.

TABLE 8
Portions of food supplying one-third of the recommended daily intake of iron (4 mg)

Curry powder	5 g (1/6 oz (1 teaspoon))
Cocoa powder	40 g (1½ oz (1 heaped tablespoon))
Baked beans	285 g (one 10½ oz tin)
Beans, peas (dry)	70 g (2½ oz (½ teacup))
Lentils (dry)	70 g (2½ oz)
Kidney	65 g (2¼ oz)
Liver, faggots	50 g (1¾ oz)
Beef (cooked/corned)	140 g (5 oz)
Black pudding	20 g (¾ oz)
Fish, fried (average)	570 g (20 oz)
Sardines/pilchards	140 g (5 oz)
Oatmeal/millet	95 g (3¼ oz)
Bran	30 g (1 oz)
Almonds	95 g (3¼ oz)
Heart	90 g (3 oz)
Soya flour	55 g (2 oz (2 tablespoons))
Shellfish (average)	35 g (1¼ oz)
Treacle	40 g (1½ oz (1 tablespoon))
Bread, wholemeal	160 g (5½ oz)
white	230 g (8 oz)
Hovis	90 g (3 oz)
Wheatgerm	40 g (1½ oz)
Greens (cooked/raw)	340 g (12 oz)
Liquorice allsorts	60 g (2 oz)
Yam/potatoes (boiled)	1 300 g (45 oz)
Flour, white	165 g (6 oz)
wholemeal	100 g (3½ oz)
Figs and apricots (dry)	85 g (3 oz)
Prunes (dry) with stones	170 g (6 oz)
Pasta (dry)	300 g (10 oz)

IRON SUPPLIES

Iron-deficiency anaemia is common all over the world, both in the 'have' and 'have-not' countries. Women and elderly people are especially affected, but men in the tropics also suffer from anaemia because of parasitic infection. Babies and teenagers, during their stages of rapid growth, can all too easily become anaemic. It is, therefore, important to provide iron-rich foods in the diet at all times.

In Great Britain about one-third of the iron in the diet comes from cereals; this is mainly because white flour is enriched with iron to the extent of 1·65 mg per 100 g, so that bread and all flour products contain added iron. Another third of the iron comes from meat, and vegetables account for one-sixth. To make sure that there is enough iron in the diet select a group of foods that contain ½ to ⅔ of the daily intake, i.e. about 6–7 mg, and can be eaten daily. In addition, frequent use should be made of meat offals, shellfish, curry, and other foods that are rich in iron.

Suggestions for planning a daily ration supplying 6–7 mg of iron (using portions list)

- 80 g (2½ oz) wholemeal bread
- 115 g (4 oz) white bread
- 40 g (1½ oz) breakfast cereal
- 10 g (⅓ oz) cocoa or meat extract

- 40 g (1½ oz) oatmeal (as porridge, oatcake)
- 15 g (½ oz) bran
- 230 g (8 oz) white bread

- 70 g (2½ oz) pulses (dry)
- 660 g (22 oz) yam or potato (boiled)

- 100 g (3½ oz) wholemeal flour (as chupattis)
- 5 g (⅕ oz) curry powder

- 115 g (4 oz) white bread
- 20 g (¾ oz) wheatgerm (with cereal)
- 70 g (2½ oz) beef, cooked

- 90 g (3 oz) Hovis
- 15 g (½ oz) bran
- 10 g (⅓ oz) cocoa or meat extract

plus 115 g (4 oz) liver or soya flour weekly

Menus that provide good supplies of iron could include seafood *hors-d'œuvres*, dishes such as braised kidneys, grilled steaks, steak and kidney pie, pasties with corned beef, curried mince, meatball curry, hamburgers with minced ox-liver mixed in. Puddings could consist of chocolate gateaux (using cocoa and almonds), stewed figs, apricot charlotte made from dried fruit, flapjacks made with oatmeal and treacle and butter. Useful snacks are sandwiches made with sardines or liver or curry paste, gingerbread, and hot chocolate as a drink.

Although it has long been known that zinc is part of several enzymes it was only in 1961 that zinc deficiency was observed in man. The signs are anaemia, and delayed sexual maturity in young men, but such cases are rare.

VITAMIN C TO AID ABSORPTION

A list of foods rich in vitamin C is given later; but the following are some suggestions for serving foods rich in both iron and vitamin C.

(1) **Curries of meat, eggs, fish, vegetables, fruit, pulses**—served with 'sambals' such as salad of sliced lemons, onion, parsley, sugar and

oil; or a salad of sliced orange, tomato and onion with chives and French dressing; or a sliced green pepper vinaigrette.

(2) **Mixed grills**—served with grilled tomatoes or tomato juice and broccoli or green salad of watercress dressed with oil and lemon juice.

*(3) **Paella**—with shellfish, peppers and tomato.

(4) **Figs**—soaked in orange juice.

*(5) **Gingerbread**—served with a glass of grapefruit juice as a snack.

(6) **Prawn cocktail**—with tomato juice and lemon.

(7) **Roast Beef**—and horseradish mixed with salted cream.

* Recipes see p. 193 Appendix.

THINGS TO DO:

1. Weigh out and make an exhibit of portions of food supplying 4 mg iron.
2. Using the portions, make up an attractive dish—for one, two, or three.
3. Make menus for two meals; (a) rich in iron (b) poor in iron.
4. Take a recipe for chocolate cake or gingerbread; calculate how much iron each portion supplies.
5. Make suggestions for increasing the iron content of the recipe and for adding the vitamin C when it is served.
6. Using a suitable scale, make a graph of the recommended dietary allowances.
7. Make a list of average portions eaten of these foods, and calculate the amount of iron provided: liver, bread, corned beef, eggs, frozen peas, porridge and potatoes, sweet potatoes, rice, greens, curry powder.
8. Make a block diagram, showing what proportion of the daily dietary allowance each portion provides.
9. Make a poster suitable for a school or works canteen to interest people in choosing the best foods for vitality.

Part 2 VITAMIN C (ASCORBIC ACID)

It has already been said that vitamin C helps the absorption of iron from the intestine but this is only one of its functions. Its main function is to assist in the formation of the connective tissue, the cement between the cells, in all parts of the body. When there is a shortage of vitamin C the disease that results is scurvy.

SCURVY

The characteristic feature of this disease is that the connective tissue between the walls of the cells in various tissues of the body begins to break down. When this happens in the walls of the fine blood capillaries small patches of haemorrhage appear under the skin around the openings of the hair follicles (these are called petechiae). Later, large, spontaneous bruises appear. The patient becomes feeble and listless. The gums become soft and spongy and the teeth become loose and eventually fall out.

In infants the first signs of bleeding show in the area immediately over the long bones and there is considerable pain.

As connective tissue depends on vitamin C for its formation, a shortage of the vitamin causes delay in the healing of wounds and there may be a breakdown of old scar tissue.

Scurvy is the end result of a long period on a diet grossly deficient in vitamin C but what is not so clear is the result of a borderline intake of the vitamin. We know that the amount consumed in northern climates falls in winter as fruit and vegetables, the richest sources of the vitamin, become scarce and expensive, and potatoes, another important source, have been long in store and lost some of their vitamin C. We also know that infections, lassitude and various disorders are more common at this time of the year. While we cannot say that these effects are a result of a shortage of vitamin C it is probably a reasonable insurance policy to make sure that the intake of this vitamin is kept as high as possible in winter time.

Scurvy is an old disease and was common in Britain in winter until the eighteenth century when potatoes became a common food. The increased production of vegetables by Danish market gardeners in Queen Anne's time helped too. Before that time people used such anti-scorbutics as sorrell, watercress, onion tops, turnip tops and nettles—all sources of vitamin C. They were boiled or chopped with vinegar as 'sallets'. Scurvy grass, which grows on the seashore, was gathered by children and steeped in ale. In Siberia and northern Europe people made decoctions of pine needles and dried rosehips, and used sprouted grains and pulses; they still do.

Seeds such as grain and pulses do not contain any vitamin C but when they are allowed to sprout they form the vitamin. An Indian recipe uses unsplit dhal or gram, about 1½ to 2 oz (42–60 g) per person, soaked in water for 12 to 24 hours. Then the water is poured off and the grain kept damp with access to air with a damp blanket. After a few hours they begin to sprout and the process is complete when the sprouts are ½ to 1 inch (1–2 cm) long. This may take about 30 hours and the maximum vitamin C content reaches 9 to 15 mg per oz (30–50 mg per 100 g).

BRITISH 'LIMEYS'

Scurvy was common among sailors in the sixteenth to eighteenth centuries because vegetables and fruits were rarely available when they were away from land. When Vasco de Gama sailed round the Cape of Good Hope he lost 100 out of his crew of 160 from scurvy. It was known as early as 1535 by the French explorer Jacques Cartier, that the juice of the leaves of a certain tree, probably the spruce fir, was curative but the real evidence came from an experiment by Dr. James Lind, a British naval surgeon. He showed that of the various remedies of the day, only oranges and lemons were effective. Captain Cook demonstrated that a long journey could be undertaken without deaths from scurvy—he was 3 years on his voyage round the world—by taking on supplies of fruit and vegetables at every possible point.

From 1792 all British sailors were given a daily ration of lemon juice (limón) and scurvy was abolished—since that time British sailors have been called 'limeys'.

LIKELY VICTIMS

Today acute scurvy is rarely seen in this country but it does occur from time to time. There were outbreaks during the First World War when it was nicknamed Bachelors' disease, because bachelors living alone could not be bothered to prepare potatoes and vegetables and fell victims to scurvy—and it still occurs in these circumstances. During the Second War efforts were made to inform the public of sources of the vitamin and methods of preparing vegetables so as to avoid destruction of the vitamin. In addition, orange juice was provided for expectant mothers and children up to 5 years of age.

There are still two groups of people who are vulnerable in this respect, babies and elderly people. Infantile scurvy occurs in well fed communities from time to time and 79 cases were reported in one town in Canada in 1958. Babies need a special source of the vitamin as their usual foods like milk and cereals contain little or none. That is why orange juice, rosehip syrup or blackcurrant juice is a necessary addition to the diet of infants. Elderly people who live largely on starchy foods with too little fruit and vegetables may also suffer from a shortage of vitamin C.

ASCORBIC ACID

As a shortage of vitamin C leads to scurvy it has been called the anti-scorbutic vitamin or, as it is chemically an acid, ascorbic acid. It occurs mainly in fruits and vegetables and only small, unimportant amounts are found in freshly killed meat and milk that has not been exposed to light. In bright sunshine a bottle of milk left on the doorstep can lose all its ascorbic acid in about 20 min. and even on a dull day it all goes in a couple of hours. But as milk is such a poor supplier of the vitamin anyway, this loss is not important.

It is an astonishing fact that only a few species of animals, including man, monkeys and guinea pigs, need vitamin C in their food. Almost all other animals can synthesize it for themselves from glucose. This synthesizing system is absent from man and so ascorbic acid is a dietary essential.

HOW MUCH DO WE NEED?

While we do not know with any accuracy the requirements of an individual, we do know the average needs of groups of people. The average plus an extra allowance for those with above average needs should provide enough for everyone. This figure is the recommended daily intake (R.D.I.).

With regard to vitamin C there has been, for many years, a discrepancy between figures given in tables in U.K. and U.S.A.; the former recommend 30 mg per day, the latter originally recommended 70 mg, in 1968 reduced the figure to 60 mg, then in 1973 to 45 mg.

The difference arose because the U.S.A. authorities based their earlier figures on the amount that would be needed to keep human blood levels of vitamin C as high as it is in animals that make their own vitamin C. The British figures were derived from direct experiment. Human volunteers who lived on a diet completely free from vitamin C (the Sheffield experiment of 1944–45) developed scurvy after 6 months. They were cured with a daily dose of 10 mg but needed 20 mg to promote wound healing. The figure of 30 mg was recommended as a safety factor. The Food and Agriculture Organization supports this figure.

The vitamin C content of raw and especially cooked foods can vary enormously, as described later, so that food composition tables are a poor guide in this respect. For this reason it may be a safety precaution to aim higher than the R.D.I., i.e. at a daily intake of 50 mg. Any surplus is harmless and is excreted in the urine in a few hours.

THE BEST FOODS FOR VITAMIN C

Fruits and vegetables are the main sources of vitamin C and few other foods make any significant contribution to the intake. There are traces of the vitamin in raw meat, fish and small amounts in milk, but these make little contribution unless people eat large amounts of fish, as do the Eskimos, or very fresh meat, like the Masai who live largely on milk, meat and blood from their cattle.

The ascorbic acid content of foods varies with species, soil, climate, ripeness and storage, but average figures for the raw foods, like average prices, are helpful in planning menus. These values may be considerably reduced by losses in preparation. There may be a loss of as much as 50% on cooking root vegetables and 70% on cooking greens. Losses on cooking fruits are usually somewhat less because their acidity protects the vitamin from oxidation. The main loss of vitamin C, however, is by dissolving out into the cooking water.

TABLE 9
Sources of vitamin C
(Raw, edible portion, unless otherwise stated)

	mg per oz	*mg per* 100 *g*	*Portions supplying* 25 *mg*
Apples, eating, raw	1	4	1½ lb (700 g)—3 large apples
Bananas	3	10	9 oz (250 g)
Bean sprouts	8	30	3 oz (85 g)
Beetroot, cooked	2	6	¾ lb (340 g)—1 large
Blackcurrants, raw	55	200	½ oz (14 g)—½ tablespoon
cooked or canned	40	140	¾ oz (20 g)—1 tablespoon
Breadfruit	6	20	4 oz (114 g)
Cabbage, cauliflower, raw,	18	60	1½ oz (42 g)—teacupful
cooked 5 min	6	20	5 oz (140 g)—¼ pint volume
cooked, kept hot 30 min	3	10	8 oz (230 g)—½ pint
Gooseberries	12	40	2 oz (57 g)
Grapefruit juice, canned	12	40	2 oz (57 ml)—wineglassful
Grapefruit, whole	5	17	5 oz (140 g)—½ small
peeled, fresh	12	40	2 oz (57 g)—8 sections
Guava, canned	50	180	½ oz (14 g)
Horseradish	34	120	¾ oz (20 g)
Kale, broccoli, beet greens, turnip tops	34	120	¾ oz (20 g)

	mg per oz	*mg per 100 g*	*Portions supplying 25 mg*
Leeks	6	20	4 oz (114 g)
Lemons, whole	23	80	1 oz (30 g)
juice	14	50	2 oz (57 g)—wineglassful
Lettuce	4	14	6 oz (170 g)—1 large
Mango	8	30	3 oz (85 g)
Melon	8	30	3 oz (85 g)
Mustard and cress	11	40	2 oz (57 g)
Okra	7	24	3½ oz (100 g)
Oranges, whole	10	35	2½ oz (70 g)—½ small orange
peeled	15	50	1½ oz (42 g)—4–5 sections
canned juice	12	42	2 oz (57 g)—wineglassful
Parsley	44	150	½ oz (14 g)
Paw-paw	14	50	2 oz (57 g)
Peas, raw	7	24	3½ oz (100 g)
fresh, cooked	4	15	6 oz (170 g)
sun-dried, cooked	0	0	
canned, cooked	3	10	9 oz (270 g)
dried by modern methods, cooked	3	10	9 oz (270 g)
Peppers, green or red	30	100	¾ oz (20 g)—piece 3 × 1¾ in
Pineapple, fresh	7	24	3½ oz (100 g)—1 slice ½ in thick
canned juice/fruit	3	10	10 oz (280 g)—½ pint (small tin)
Plantain	6	21	4 oz (114 g)
Potatoes, new, cooked	5	18	5 oz (140 g)—2 med. size
old, after January	1	4	1½ lb (700 g)
Redcurrants	12	40	2 oz (57 g)
Rosehip syrup	56	200	½ oz (14 g)—2 teaspoons
Spring onions	7	25	3½ oz (100 g)
Sprouts, raw	28	100	1 oz (28 g)—2–3 sprouts
cooked	10	35	2½ oz (70 g)—5–6 sprouts
Strawberries	17	60	1½ oz (42 g)
Swedes, cooked	5	18	5 oz (140 g)—large portion
Sweet potato	8	28	3 oz (85 g)
Tomato, raw, cooked or juice	7	25	3–4 oz (100 g)—1–2 large tomatoes
Watercress	18	60	1½ oz (42 g)
West Indian cherry	570	2 000	
Yams, fresh	3	10	9 oz (270 g)

Note: The vitamin content of foods, and especially vitamin C, varies considerably from one sample to another, and the above figures can be used only as a guide.

There are some fruits and vegetables, not given in Table 9, that contain only 3 to 10 mg of vitamin C per 100 g or 1 to 3 mg per oz. They include pears, parsnips, marrow, aubergine, pumpkin, endive, chicory, plum, peach, apricot, French beans, fresh figs, grapes and cucumber. If they are eaten in large amounts, as for example yams and bananas are, then they can make a substantial contribution to the vitamin C intake.

The best example of quantity effect is the potato. In terms of vitamin content per ounce they cannot be termed a rich source and old potatoes, at only 1 mg per oz (3–4 mg per 100 g), are a very poor source, but as so much potato is eaten we find that, on average, it provides one-third of our vitamin C intake in Great Britain in summer time, and as much as half the total in winter time when much less fresh fruit is eaten.

POTATOES

The ascorbic acid content of potatoes is highest when freshly dug; they average 9 mg per oz (30 mg per 100 g). During storage the ascorbic acid content steadily falls until by the end of the winter it may be as little as 1 to 2 mg per oz.

TABLE 10

Vitamin C content of potatoes

Month	*Raw*	*Boiled*	*25 mg portion*
	per oz		
New potatoes	9 mg	5 mg	5 oz (140 g)
October–December	5 mg	3 mg	8 oz (230 g)
January–February	3 mg	2 mg	¾ lb (340 g)
March–April	2 mg	1 mg	1½ lb (700 g)

Note: (1) The daily requirements of vitamin C can be met by serving large-average portions of potatoes, yams or bananas, or smaller portions of sweet potatoes or plantains.

(2) The cheapest source of vitamin C in Great Britain is likely to be raw cabbage from June onward. Rosehip syrup is a cheap source, too.

Example of calculation: lettuce contains 4 mg per oz (14 mg per 100 g), therefore 25 mg will be present in 25/4 = 6 oz (170 g). Lettuce has an average waste of 20%, i.e. 80% is edible. To find how much to buy in order to get the 6 oz portion divide 6 oz by 80% (= 7½ oz (215 g)). The *cheapest* sources of vitamin C are found by comparing the cost of the portions; or by calculating the amount of vitamin C per penny (Appendix p. 188).

HOW TO ENSURE AN ADEQUATE SUPPLY

(1) The easiest and most practical way is to choose a vitamin C-rich food which will supply the whole day's need in an average portion, e.g. one small orange, or 4 oz (114 g) portion of grapefruit or 4 oz (114 ml) of orange juice.

(2) The traditional way in Great Britain is to use a portion of potatoes (8 oz—230 g) and a portion of swede or greens (5 oz—140 g). This plan breaks down (a) if the food is kept hot before being served, (b) if the potatoes are mashed because they lose the vitamin faster than do the whole potatoes, and (c) in late winter when the vitamin C content of potatoes is low.

(3) The cheapest way is to serve a 3 oz portion of cole slaw (raw cabbage salad). Synthetic vitamin C in powder or tablet form is also a very cheap way of providing the vitamin. It is used to enrich drinks, fruit juice, etc., and also on expeditions.

(4) Further suggestions—use any two of the following:

Breakfast: 3½ oz (100 g) portion of tomatoes, fried, grilled or canned (with bacon), half a grapefruit, or prunes soaked in 2 oz (57 ml) of canned orange juice.

Main meal: Prawn cocktail using 2½ oz (70 ml) of tomato juice, ¼ oz (7 g) chopped green pepper.

Florida cocktail—1 oz (28 g) each of orange and grapefruit segments.

Serving of green salad, including ¾ oz (20 g) watercress and 1 oz (28 g) spring onion.

Serving of fruit salad, including 1 oz (28 g) each of orange, melon and fresh pineapple.

Portion of blackcurrant and apple pie (1 oz (28 g) blackcurrants) or 1½ oz (42 g) strawberries (fresh or frozen) and ice-cream.

AVOIDING LOSSES

Vitamin C is the vitamin most sensitive to losses in food preparation and, therefore, one to which most attention must be paid.

To make sure of getting enough vitamin C the first step is to make a good choice of greengrocery. Many fruits and vegetables, as shown in the table, are poor sources even when fresh. The second step is to store and prepare them in such a way as to avoid loss.

1. VITAMIN C IS EASILY DISSOLVED IN WATER

When fruits and vegetables are cooked a large part of the vitamin C (and this applies also to the water-soluble B vitamins) can be washed out into the cooking water. Losses are greater the more finely the food is chopped and the more water is used in cooking. For example, a leafy vegetable completely immersed in water can lose up to 80% of its ascorbic acid, but if ¼ covered the loss is only 40%. If the water is consumed, as is usual in cooked fruits, or if the vegetable water is used immediately for making gravy, then the vitamin is not lost.

Recommendations

A. Soak the vegetables in water for only a short period or wash them using brush and spray.

B. Cook them in the minimum amount of water in a closed pan.

C. Reduce both cooking time and contact with water by shredding leaves (not too finely) and cutting roots into small pieces. Pressure cooking is a good method of cooking without much loss of the vitamin.

D. Losses are reduced by avoiding water, i.e. cooking in fat. For example chipped potatoes (23% loss), roast, peeled, cut potatoes (58% loss), potatoes baked or boiled in jackets (20% loss), Chinese stir-fry vegetables.

2. VITAMIN C IS EASILY OXIDIZED

Vitamin C may be oxidized both by air and by an enzyme, ascorbic acid oxidase, in the vegetable itself. Air oxidation means that if the vegetable is kept hot there is a rapid destruction of the vitamin. For example mashed potatoes kept hot can lose half their vitamin content in 20 min.

In the living vegetable the enzyme, ascorbic acid oxidase, is separated from the vitamin, but when the tissue is damaged by wilting, bruising, freezing, grating or chopping, the enzyme is brought into contact with the vitamin and destruction begins.

Recommendations

A. Buy the greengrocery fresh and handle it gently to avoid bruising. Protect it from the frost and do not allow it to become frozen in the refrigerator; keep it in the cool rather than the cold part. Keep green leafy vegetables cool and moist in closed plastic bins (not zinc, see later) or closed polythene bags in the cold. The wire racks so often sold for storing fruits and vegetables are a poor means of storage.

B. The enzyme is rapidly destroyed in boiling water so minimize the loss from this cause by putting the vegetables, a handful at a time, into boiling water. Reboil after each addition. Use small, quick-boiling pans.

C. Eat the cooked vegetables as soon as possible after cooking.

3. METALS SPEED UP THE OXIDATION OF VITAMIN C

Traces of metals, such as iron, zinc and, most especially, copper speed up the oxidation of the vitamin. Sufficient will dissolve from knives and cooking utensils to make a marked difference to the amount of vitamin C left in the food.

Recommendations

A. Use plastic vessels for the storage and preparation of fruits and vegetables not zinc. Enamel vessels are all right until the enamel begins to chip off. Equipment such as knives, graters, sieves, colanders, and strainers should be of stainless steel, aluminium or plastic. Liquidizers and mincers should have cutters of stainless steel, special alloy or plastic.

B. Copper cooking vessels will lead to rapid destruction of the vitamin; pans and boilers should be made of stainless steel, aluminium, fireproof glass or unchipped enamel. Any iron or tinned iron vessels should be reserved for foods such as carrots, beetroot, parsnips, etc., which have little vitamin C to lose.

4. HEAT DESTROYS VITAMIN C (IN AIR)

Raw fruit and vegetables are always richer in vitamin C than the cooked ones, but if they are to be cooked then short, rapid heating is much less destructive than long, slow heating. Keeping hot and reheating are very destructive. Losses on keeping vegetables hot vary with the type of vegetable and the surface area but roughly a quarter would be destroyed by keeping hot for 15 min., half in 45 min. and 80 to 90% in one hour.

Recommendations

A. Organize vegetable cooking so that they can be cooked quickly and served at once on a hot dish. Do *not* keep vitamin C-rich vegetables hot in insulated containers or hot cupboards.

B. Do not assume that any significant amount of vitamin C is left in the vegetables in soups, stews and braises.

C. In large-scale catering, where vegetables are cooked in large boilers, and kept hot for even a quarter of an hour, there can be considerable losses of vitamin C. A better method is to cook vegetables in a series of small boilers of the quick-heating type, and to keep up a continuous service of freshly cooked vegetables.

D. Mashing and puréeing of cooked vegetables and fruits, such as mashed potatoes, puréed spinach and gooseberry fool, improve flavour and interest but reduce the vitamin C content. Therefore, when this is done, some other source of vitamin C should be added to balance the menu.

E. Reheated leftover vegetables such as bubble and squeak and colcannon, will have lost all their vitamin C. It could be replaced by adding tomatoes or mixing in freshly cooked swede.

F. Convenience foods. Canned fruits and vegetables may lose about a quarter of the vitamin C during manufacture but the rest will keep for many months in the sealed can. A little more will be lost on heating the contents of the can. If this is a short heating period the final result is not very different from cooked fresh fruit and vegetables.

Dried foods are poorer than canned ones in vitamin C unless they have been dried under vacuum or by the new accelerated-freeze-drying process.

Frozen foods retain most of their original vitamin C content but lose it rapidly on thawing.

Bottled fruit juices will keep their vitamin C content for several months in a full, closed bottle but can lose it all in a week once the air gets into the opened bottle. Keeping it cold will help to preserve the vitamin a little better.

5. EFFECTS OF COLD

At cool temperatures, i.e. above freezing point, plant activity such as ripening and growth is checked and fruits and vegetables do not wilt, if they are protected from drying.

Slow freezing, such as occurs in frost or in the refrigerator below freezing point, forms ice-crystals which damage the tissues and then vitamin C is lost through coming into contact with the enzyme, ascorbic acid oxidase.

Quick-freezing at temperatures of −10 °C or −27 °C prevents the formation of large ice-crystals and the plant tissues are undamaged so the vitamin C is not lost.

Enzymic destruction is prevented by a preliminary blanching in boiling water which destroys the ascorbic acid oxidase.

Retention of vitamin C in frozen foods

At −27 °C, 1 year, vegetables—no loss.
At −18 °C, 4 months, fruit juices—no loss.
At −18 °C, 8 months, peas, broccoli—15% loss.
At −18 °C, 8 months, cauliflower, spinach—50% loss.
At −12 °C, 4 months, vegetables and fruit juices—50% loss.

Recommendations

A. Fruits and vegetables should be stored in a cool place but not below freezing point. Leafy vegetables should be protected from drying by storing in closed plastic boxes or polythene bags.

B. Deep-frozen foods are often cheaper than fresh ones because of lower preparation costs and less wastage. Care should be taken that they have not been allowed to thaw out and then refrozen.

6. EFFECTS OF ACIDS AND ALKALIS

Vitamin C, itself an acid, is stable to acids but destroyed by alkalis.

Recommendations

A. Vinegar, citric acid or lemon juice helps to preserve the vitamin, but cannot be used for green leaves because it yellows the chlorophyll. Acidified boiling water should be used to blanch vegetables for deep freezing. Acid pickles retain much of the vitamin as, for example, green walnuts, marinated sweet peppers, horseradish, piccalilli, red cabbage in vinegar and sauerkraut. French dressing reduces the loss in salads.

B. Bicarbonate of soda is often added to the cooking water to improve the colour of greens. An excess will destroy the vitamin C and it is not necessary if the greens are fed slowly into fast-boiling water and cooked quickly. A *small* pinch of soda is not harmful and if it helps to make the vegetables more attractive its use may be justified.

HOW IMPORTANT ARE COOKING LOSSES?

If the diet includes a rich source of vitamin C in a food that is not cooked, such as fruit juice, raw fruit or raw vegetables, then cooking losses in other foods do not matter. But in schools, restaurants and industrial canteens where people may prefer cooked vegetables and may rely on the main meal for their vitamin C supplies, then the caterer must make sure that even after cooking the foods supply half the daily recommended allowance, i.e. about 25 mg. If old equipment and inadequate facilities allow too great a loss of the vitamin then other sources must be found and popularized. The obvious sources are fruit juices and concentrates and raw vegetables.

(1) GREEN SALADS

(a) 1 oz (28 g) lettuce, $\frac{1}{3}$ oz (10 g) watercress, $\frac{1}{4}$ oz (7 g) chopped green pepper—21 mg vitamin C.

(b) 1 oz (28 g) cabbage, $\frac{1}{2}$ oz (14 ml) lemon juice—24 mg vitamin C.

(c) 1 oz (28 g) lettuce, 1/8 oz (3 g) parsley—9 mg vitamin C.

(2) MIXED SALADS
(a) ½ oz (14 g) lettuce, ½ oz (14 g) spring onion, 1½ oz (42 g) tomato—15 mg vitamin C.
(b) 1 oz (28 g) chicory, ½ oz (14 g) spanish onion, 1 oz (28 g) tomato, 1 oz (28 g) orange—25 mg vitamin C.
(c) ½ oz (14 g) lettuce, 1 oz (28 g) cucumber, 1 oz (28 g) cooked beet—6 mg vitamin C.
(d) 1 oz (28 g) cooked beet, 1 oz (28 g) apple, 1 oz (28 g) orange, ¼ oz (7 g) spring onion—22 mg vitamin C.

(3) HORS-D'ŒUVRES
Sliced tomato and cucumber with oil, lemon and sugar dressing. Herring salad with potatoes, beet, apple, onion and orange. Orange and tomato salad with onion. Marinated red and green peppers, or sliced tomatoes with French dressing, or radishes.

(4) FRUIT JUICES
These can be drunk hot or cold, straight or mixed with mineral water, or mixed into jellies or trifles. Concentrates can be used directly as syrups poured over pancakes, ice-cream, babas and sponges, or to make sorbets or to enrich fruit salads or stewed or canned fruit.

(5) NEW WAYS WITH FOODS
Juicers and liquidizers allow fruits and vegetables to be served in new and attractive ways. They replace and improve upon mortars, tammy cloths, mincers and sieves for preparing purées and juices.

CONCLUSIONS

Ascorbic acid is an important nutrient that may sometimes be deficient in diets today, in spite of the availability of good and cheap foods. A knowledge of the vitamin content of fruits and vegetables is needed to make a good choice and a knowledge of the preparation losses that can occur is necessary to make sure that the vitamin reaches the consumer. (See p. 141.)

Poorly selected menus will be short of vitamin C and so will poorly cooked foods and especially foods that have been kept hot for any length of time or have been reheated. For this reason, a measure of the vitamin C content of a meal serves as a rapid means of assessing the 'quality' of that meal.

MEASUREMENT OF VITAMIN C

Vitamin C has the property of decolourizing a blue dye, indophenol (full name dichlorophenolindophenol) and the amount that is decolourized is a measure of the amount of vitamin C present.

1. The vitamin must be extracted from the food unless it is already in solution as in a fruit juice. This is done by grinding the food in a mortar with a little clean sand or homogenizing in a liquidizer. To prevent loss of the vitamin during the extraction, a 5% solution of acetic acid or freshly prepared metaphosphoric acid is added; this destroys the enzyme, ascorbic acid oxidase, which would otherwise attack the vitamin C.

Take 5 grammes of leafy vegetable or potato and grind or homogenize it with 50 ml acid until free from lumps. The vitamin is now in solution. Pour the mixture into a measuring cylinder and add water to give a total volume of 100 ml; shake the mixture. Use 10 ml of the upper, clearer part of the extract for the estimation—it may not need filtering.

2. The indophenol dye can be purchased from the British Drug Houses Ltd in tablets, each of which has a titration value equivalent to 1 mg of vitamin C. Dissolve the tablet in 50 ml of distilled water and fill the solution into a burette. Each ml is equivalent to 1/50 mg of the vitamin.

3. Pipette 10 ml of the food extract (it need not be clear) into a flask and add the dye in drops from the burette. The blue dye turns pink as soon as it meets the acid in the flask and is immediately bleached by the vitamin present. Continue to add the dye until a faint pink colour persists for at least 10 seconds—this means that enough dye has been added to react with all the ascorbic acid. The amount of the vitamin in the food is calculated from the amount of the dye added:

If x ml of dye was needed, then the 10 ml sample of food extract used for the titration contained $x \times 1/50$ mg of vitamin C.

The whole 100 ml of extract, i.e. the 5 g food sample, contained $10 \times x \times 1/50$ mg of vitamin C. This may be expressed in mg per 100 g or per oz (28 g).

PRACTICAL WORK

1. Measure the amount of vitamin C in freshly squeezed lemon or orange juice, orange squash, cabbage (outer leaves and inner leaves), milk.
2. Boil a sample of cabbage in small, medium and large amounts of water and measure the amount of vitamin C that is left in the cabbage and the amount leached out into the cooking water.
3. Weigh out and display portions of food containing half the daily

allowance. Arrange them in order of (a) the richest sources, (b) the cheapest.

4. Make a salad plate supplying 50 mg of ascorbic acid.
5. Prepare a drink, hot or cold, providing 100 mg. For whom would it be suitable?
6. How could you improve the vitamin C value of a winter salad consisting of cooked potatoes, beetroot, celery, onion and mayonnaise?
7. How would you add vitamin C to the following meals: (a) Ox-tail soup, roast meat, mashed potatoes, carrots, rice pudding. (b) Grilled sole, potato straws, spinach purée, chocolate gateau and cream?
8. Describe how you could provide useful amounts of ascorbic acid in (a) a milk bar, (b) sandwich bar, (c) licensed bar.

EXAMINATION QUESTIONS

1. What contribution is made by fruits and vegetables to the nutritive value of human diets?
   Which fruits and vegetables constitute good sources of the important nutrients? (R.S.H.)
2. What procedures are important in the cooking of green vegetables on a large scale? Give reasons.
   How would you use cooked potato, cabbage, carrots and peas left over after main meals? What would be their nutritive value? (R.S.H.)
3. For what reasons should vitamin C be included in the daily diet? How would you ensure an adequate intake in the dietary of hospital patients on a full diet? (R.S.H.)
4. How, in large-scale catering, would you purchase, store, prepare, cook and serve foods containing ascorbic acid to ensure conservation of the vitamin? (R.S.H.)
5. Make a list of fruits and vegetables which are poor value for vitamin C. Describe ways of serving them which improve their vitamin C value.
6. Give the main sources of iron in your own diet, and indicate how you would ensure that your daily intake is sufficient to cover the recommended allowances. Explain the function of this nutrient. (Hotel and Catering, Intermediate.)
7. Plan six packed lunches for the winter months suitable for a young typist. Give quantities of foods and calculate the iron and vitamin C available in one meal. (R.S.H.)

8. What foods are important sources of vitamin C in different age groups? Discuss the factors which affect the vitamin C content of foods as eaten. (R.S.H.)
9. Why and how would you encourage the use of citrus fruits and salads in March in the following catering establishments: (a) a licensed snack bar, (b) a milk bar, (c) a college refectory, (d) a high-class restaurant? (Battersea Hotel Management Certificate.)
10. What are the consequences of a low intake of iron and calcium* in the diets of adolescents? What foods are good sources of these nutrients? Make out a bill of fare for a snack bar, catering for students. Underline items that provide good supplies of iron and calcium.
11. Discuss the nutritional importance of salads in our National Diet. How would you popularize them in an industrial canteen, school meal?
12. Plan a short 'pep' talk to the women staff of a factory on the dangers and causes of anaemia. Show them the importance of taking enough of the right foods to prevent this illness, and suggest ways of serving these foods which will appeal to them. (R.S.H.)
13. Write a label for a bottle of vitamin C-rich fruit syrup, giving instructions for its use and dosage. The syrup contains 25 mg per oz. Give three recipes which could be used on a leaflet making good use of the syrup suitable for children.
14. What is the value in the diet of a daily supply of vegetables? Give the method and reasons for any special care needed in the preparation and cooking of (a) new potatoes, (b) cabbage. (City & Guilds 243.)
15. Prepare and serve a buffet-style lunch for sixth-formers. Include two drinks and some home-made bread. In this type of meal vitamin C and iron are often not well supplied, so provide dishes which contain significant amounts of these 2 nutrients. Calculate the vitamin C and iron content of *two* of the dishes ('A' level practical paper in Home Economics, 1972).

* See p. 73.

4 Foods for construction and repair

The body is composed of a bony framework, the skeleton, clothed with flesh. The skeleton is mainly calcium phosphate and the flesh is protein—therefore protein, calcium and phosphorus must be included in the diet.

Part 1 PROTEIN

RENEWAL OF TISSUES

In the growing child it is obvious that both muscles and bones are increasing in size so there must be a regular intake of foods to supply these materials for their construction. When the child becomes an adult, however, obvious growth stops—he may become heavier by getting fat, but he does not grow any taller—there is no further increase in the amount of bone and protein. It might be thought that there is no need for protein in the diet after growth has ceased. The adult body, however, is not static like a brick building which, once built, is a finished, permanent structure, but is continually being renovated. There is a steady process of renewal as tissues are continually being removed and replaced by fresh tissues. It is rather as if some of the bricks of a building were removed each day and replaced by new ones. The building would not grow any bigger, it would remain the same size and shape, but would be continually renewed.

In living tissues the process of renewal is called 'dynamic equilibrium'. It is a method by which the body is better able to withstand damage and the daily wear and tear. It explains why the adult needs a regular dietary supply of materials for construction.

When a wound is healing or wasted muscles are being renewed after illness, the growth of new tissue in the adult is readily seen. Apart from these examples it is not easy to see that new tissues are being made. The process was detected only when proteins and minerals were 'labelled' with radioactive isotopes so that their role in the body could be observed.

The breakdown of old tissues and their renewal takes place more rapidly in some parts of the body than in others. The vital organs such as the heart, liver, kidneys and blood proteins are broken down at such a rate that half the organ is renewed in 10 days—they are said to have a 'half-life' of 10 days. The process is much slower in muscles, skin and bone which have a half-life of 5 to 6 months. This does not mean that the liver is completely renewed in 20 days, but that half is renewed in 10 days, half of the remainder (i.e. half of a half) in another 10 days, and so on.

PROTEIN NEEDS

About 20 to 25 grammes, i.e. three-quarters of an ounce, of body protein, calculated as dry weight, is replaced each day. As body tissues are roughly four-fifths water this corresponds to 115 g or 4 oz of tissue. The figure of 20 to 25 g is the absolute *minimum amount* of protein that must be eaten to keep healthy. To cover this requirement and also provide a margin of safety, the *daily allowance* of dietary protein is suggested as 45 to 50 g per day.

This is the allowance for a healthy person. In some illnesses there is a wasting of the muscles so during convalescence additional protein is needed in the diet to rebuild the lost tissues. Additional protein is also needed in physical training when muscles are being built up, and for children and young people who are growing.

WHAT ARE PROTEINS?

All living tissue, both plant and animal, contains protein. Just as human flesh is made of protein so is the flesh of animals, birds and fishes, as well as the seeds of plants, and bacteria, yeasts and moulds. The leaves, stems and storage organs of plants contain smaller amounts of protein.

All protein is manufactured by plants from the nitrogen of the soil and carbon dioxide of the air. Animals can obtain their protein only by eating plant protein or other animals.

Proteins differ from fats and carbohydrates—the other two main constituents of food—in containing nitrogen—approximately 16%. This provides a useful means of measuring the amount of protein in a food. The procedure, known as the Kjeldahl method, consists of boiling a sample of the food with sulphuric acid which converts all the nitrogen of the protein into ammonia, and then the ammonia is measured by distilling and titrating with acid. As proteins contain 16% of nitrogen the ammonia nitrogen multiplied by the figure 6·25 (which is 100/16) gives a measure of the protein.

To be precise not all the nitrogen in food is in the form of protein so the value, 'nitrogen multiplied by 6·25' is only an approximation and is called 'crude protein'. This is usually quite adequate for dietary calculations and is the value given in food composition tables. (A very small number of foods, of which mushrooms and other fungi are examples, contain a fair amount of non-protein nitrogen and this must be allowed for. Food composition tables usually state whether or not this has been done. The factor 5·7 is used for cereals.) If true protein is to be measured, it must first be separated from the non-protein nitrogen present in the food and then subjected to the Kjeldahl analysis.

All proteins are composed of the same basic units, the amino acids. Although there are only twenty of these they form hundreds of different proteins by linking together in different combinations and different amounts. It is similar to the way in which thousands of words of the language are formed from only twenty six letters of the alphabet linked together in different combinations. We thus find in nature many different proteins, such as those of milk, eggs, peas, liver, seeds, nuts, muscle, meat, and fish, all composed of the same basic amino acid units.

During digestion the food proteins are broken down to their constituent amino acids by the enzymes of the digestive juices. The amino acids are then absorbed into the blood stream and carried to the tissues where they are rebuilt into the particular protein pattern that is muscle, heart, lungs, skin, etc. This explains why proteins from widely different sources can be used in the body.

Of the twenty amino acids the body can make twelve if the basic materials (carbon and nitrogen) are provided in the diet. The remaining eight must be provided ready-made in the diet. They are therefore termed 'essential amino acids'. The increased requirements for growth make two more amino acids, namely arginine and histidine, essential for children (see Table 11).

QUALITY

Quality is a measure of the usefulness of a food protein for the purposes of growth and maintenance of tissues. One with an essential amino acid composition similar to that of the body tissues is clearly more useful than one with dissimilar composition.

Quality may be assessed by feeding the protein food under specified experimental conditions and measuring how much is used for synthesizing body protein. The tissues have the ability to select those amino acids that they can use; the remainder are oxidized to supply energy but, so far as protein synthesis is concerned, they are wasted. Hence the percentage that is retained in the body for the synthesis of new tissue is a measure of the usefulness of the dietary protein. This percentage is termed 'biological value'. Only the proteins of egg and human milk have a biological value of 100; meat and fish have a value of 75, bread 50. That is to say that under the specified experimental conditions 100%, 75% or 50% of the protein eaten is used for tissue synthesis.

Explanation It may be helpful to compare the body fabric with a wall of coloured bricks, red, white, blue and black, built in a given pattern. The black bricks can be manufactured on the spot, i.e. so far as delivery is concerned they are non-essential. The coloured bricks must be delivered ready-made, i.e. they are essential. To build the given pattern requires a supply of the coloured bricks *in the correct ratio.*

If a delivery consists of large numbers of red and white bricks but not enough *blue* ones then only a limited length of the patterned wall can be built. When all the blue bricks have been used up building must stop, even though there is a surplus of red and white ones. *Blue* bricks are, therefore, the limiting factor.

In the same way, amino acids supplied in the diet can be built into the particular pattern of tissue proteins only so long as there is an adequate supply of all the essential amino acids in the required proportions. The amount of protein that can be synthesized will depend upon that amino acid present in least amount, the *limiting amino acid.*

If one of the colours were missing altogether then no wall could be built. Similarly if one essential amino acid is missing from our food

TABLE 11
Amino acids

Essential	*Non-essential*
Leucine	Alanine
Isoleucine	Glycine
Lysine	Serine
Methionine	Proline
Valine	Hydroxyproline
Threonine	Cystine
Tryptophan	Tyrosine
Phenylalanine	Glutamic acid
plus, for children	Aspartic acid
Arginine	Cysteine
Histidine	

then no tissue could be formed. In practice this does not happen as very few proteins are *completely* lacking an amino acid.

The proteins of egg and human milk have all the essential amino acids present in the required proportions and so are completely usable for tissue synthesis. Under the conditions of measurement 100% of the egg or human milk protein eaten stays in the body and is used for tissue synthesis—they have a *biological value* of 100. No other proteins are quite as good. For example, cow's milk has a 25% deficiency of one of the essential amino acids, namely methionine, and so has a biological value (BV) of 75. Wheat—and therefore bread—has a 50% deficiency of another essential amino acid, lysine, and so has a BV of 50.

QUALITY AND QUANTITY

It was stated earlier that the amount of protein needed to replace the continuous tissue loss is 45 to 50 g per day. This is the amount required if the protein is of 'perfect' quality, BV 100, and can all be used by the body for tissue synthesis. If the dietary protein is not perfect then more than 45 to 50 g is needed. Quantity can make up for quality. 90 g of a protein of BV 50 is as good as 45 g of a protein of BV 100.

To return to the analogy of the brick wall—if the delivery consists of the exact number of bricks needed to build a given length of the wall, then the colours must be in the right proportions. But if the number delivered is greater than needed, even if the colours are not in the right proportions, it may still be possible to select enough of each colour to build the required length of wall. Quantity has compensated for quality.

After the body has selected the amino acids that it can use, the rest is oxidized to supply energy. When amino acids are oxidized in the body, the nitrogen fraction is excreted as urea in the urine. When a protein of low BV is eaten much of it cannot be used for the tissue synthesis and, therefore, the amount of urea in the urine is greater than when a protein of high BV is eaten; thus the urine nitrogen excretion is an index of the quality of a protein and is the basis of the biological value test.

Measurement of biological value In this test a young, growing, experimental animal (the rat is commonly used) is fed on a diet containing only 10% protein, whereas for good growth it would need about 20% protein. At this lower level *all* the protein that is fed is taken for growth and there is no surplus to be oxidized *except that part which is unusable for tissue synthesis because the quality is not good enough.* The unusable portion is oxidized and its nitrogen is excreted in the urine as urea and can be measured. In this way the percentage of the dietary protein that is *kept in the body*, i.e. used for tissue synthesis, can be measured, this is the measure of quality. (To be precise, there are two measurements involved here: the percentage of the food protein that is *digested* and the percentage of the absorbed food that is *retained* in the body. It is this last percentage that is termed biological value—the percentage of the *absorbed protein* that stays in the body. The name used for the percentage of the *protein eaten* that stays in the body (that is if the digestibility is also taken into account) is *net protein utilization* or NPU. In practice, about 90 to 95% of the protein of the normal diet is digested, so that NPU and BV are often similar in value.

COMPLEMENTATION BETWEEN DIFFERENT PROTEINS

The analogy of the brick wall has further uses. If one load is short of red bricks but has a surplus of blue ones, and another load delivered *at the same time* has a surplus of red bricks and a shortage of blue ones, then the two loads will complement each other. More wall can be built from a combination of the two loads than from the two separate deliveries.

A similar complementation takes place between proteins. If one protein is short of lysine but has a surplus of methionine, and another has a surplus of lysine but is short of methionine, then a mixture of

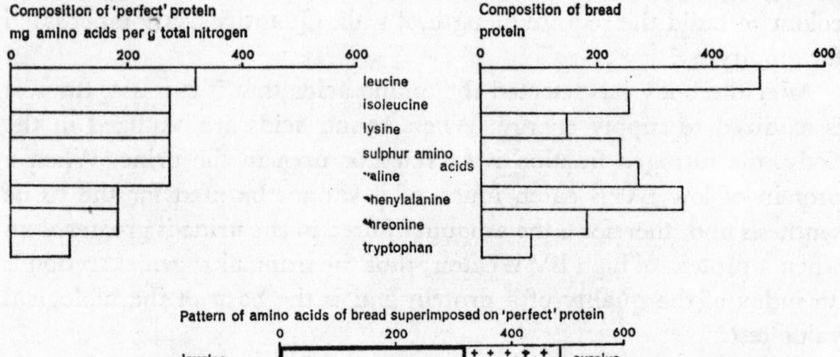

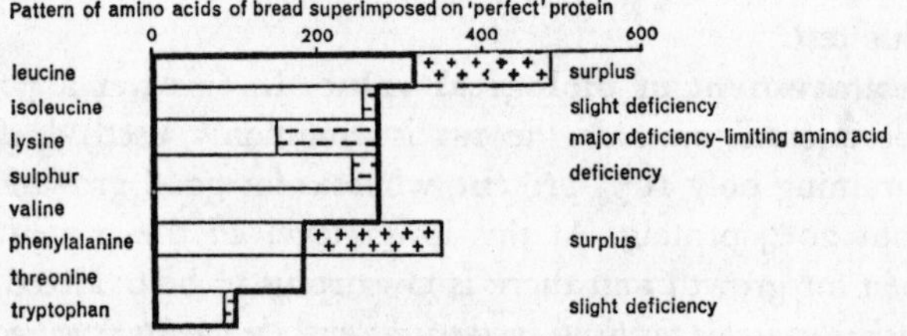

Fig. 15 Amino acid composition of bread

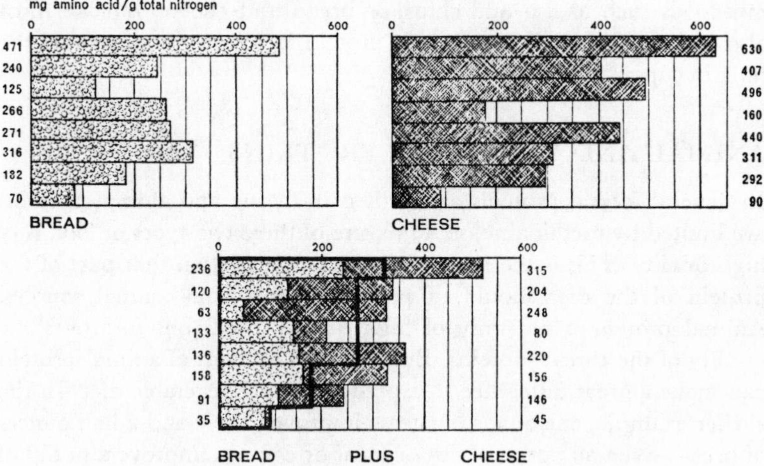

Fig 16 Complementation between amino acids of bread and cheese

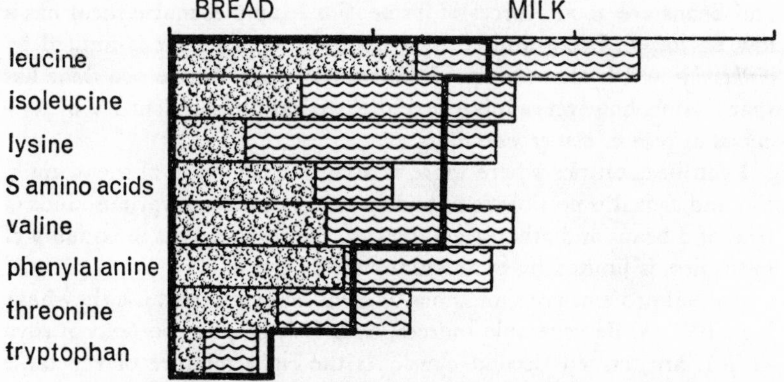

Fig. 17 Complementation between proteins. Pattern of 'perfect' protein superimposed on mixture of equal parts of bread and milk proteins

the two has a higher BV than we might expect from the average of the two separate BV's. For example, bread has a BV 50 and cheese 75; a mixture of $3\frac{1}{2}$ parts of bread and one part of cheese has a BV of 75 (the expected result would be 55). The shortage of lysine in the bread is supplied by the surplus present in cheese (see Figs. 15, 16 and 17).

The two proteins must be consumed at the same time as there is little storage of amino acid mixtures in the body. In practice, we eat mixtures of several proteins at every meal because not only do we mix foods such as fish and chips, or bread and cheese, but we must also include the small amounts of protein in soups, puddings and the milk in cups of tea.

ANIMAL AND VEGETABLE PROTEINS

In general, cereal proteins are limited by lysine and animal proteins are limited by methionine, so a mixture of these two types of food is of high quality. This is the basis of the recommendation that part of the protein of the diet should, if possible, come from animal sources. Animal protein is not only of high quality itself, but improves the quality of the cereal proteins. Even a small amount of animal protein can make a great difference to a predominantly vegetable diet. In the earlier example, one ounce of cheese improved three and a half ounces of bread. Even an ounce of fish or meat or egg can improve a potful of rice or maize.

It is possible, of course, to supplement a lysine-deficient diet with any food that is rich in this amino acid, *it need not be animal protein.* Peas and beans are good sources of lysine. For example, maize meal has a low BV of 36, being limited by *lysine*; yellow pea flour is limited by *methionine* and is almost as poor with a BV of 40. As the pea flour has spare lysine, however, a mixture of the two has the excellent BV of 70—*almost as good as that of meat at* 75.

Even in countries where there is a chronic shortage of meat, milk, fish and eggs it is possible to obtain enough lysine from various kinds of peas and beans and other vegetables; so the result is that the quality of most diets is limited by methionine, not by lysine.

Animal proteins are sometimes termed 'first class' because they have high BVs, while vegetable sources, with BVs of 40 to 60 (except soya at 70), are termed 'second class'. As the right mixture of vegetable proteins can be first class this distinction is not valid.

To make mixtures of good quality it is necessary to have some knowledge of the amino acid composition of the food proteins. These are given in food composition tables; but in the absence of detailed knowledge two general principles may be helpful. (1) Vegetable dishes should be supplemented with as much animal protein as is available (this certainly produces a very palatable dish) and (2) full use should be made of peas and beans which have a relative surplus of lysine. Proteins limited by the same amino acid do not, of course, complement one another. A mixture of equal amounts of maize at BV 36 and bread at BV 50, both limited by lysine, has a BV of 43, which is the average of the two separate values.

An example of the effect of mixing proteins is shown by comparing 4 oz (115 g) of fried steak with a portion of spaghetti bolognaise (4 oz (115 g) of spaghetti, 1 oz (28 g) stewing steak and 1 oz (28 g) cheddar cheese). The steak provides 23 grammes of protein and the spaghetti bolognaise provides 27 grammes, of nearly the same quality, *for one-third of the cost.*

SYNTHETIC AMINO ACIDS

In recent years both methionine and lysine have been synthesized and manufactured on a commercial scale. If one of these amino acids is added to a protein *in which it is limiting* there is an increase in BV. (The addition of a non-limiting amino acid has no effect.) A few manufactured foods are supplemented in this way. There are, for example, speciality loaves in the United States which contain added lysine, and some of the protein-rich mixtures prepared in the developing countries are supplemented with methionine. Amino acid supplementation is quite unnecessary in western countries where the average protein intake is 75 to 90 g per day of BV about 70 and as regards the developing countries, it may be more practicable to provide extra protein than to supplement in this way.

Amino acid supplementation may be more practicable in animal feeds. Here the question is simply one of economics, it is sometimes cheaper to improve the quality by adding the limiting amino acid, in other circumstances it is cheaper to feed extra protein.

VEGETARIAN DIETS

It follows from the previous discussion that from the point of view of proteins a purely vegetable diet need not be inferior to a mixed diet. Certainly, the usual lacto-vegetarian diets which include milk, cheese and eggs need not lack in quantity nor in any of the essential amino acids, as is illustrated by the excellent physique of the Sikhs described in Chapter 1. The *purely* vegetable diets, such as are eaten by Vegans, would need careful planning to ensure an adequate protein supply. Without such planning there could be a deficiency in both quantity and quality. The diet will be rather bulky as most vegetable foods are not as concentrated a source of protein as are the animal foods. See also p. 36.

CONCENTRATED PROTEIN FOODS

There are a number of foods which are so concentrated a source of protein that they are specially useful in supplementing protein-poor diets. Dried milk, both full-fat and skimmed, cheese and eggs provide plenty of protein in small space. An excellent vegetable source is soya flour, both full-fat and defatted. Nuts are also a good source; almonds contain 20% protein, Barcelona nuts 14% and peanuts 28%. Curd cheese contains 24% and dried yeast 50% protein.

HOW MUCH PROTEIN?

It was stated earlier that the recommended daily intake of protein is about 45 to 50 g, assuming the quantity is 'perfect'; but this figure will vary with the individual. First, the size of the body will govern the maintenance requirements, i.e. the amount needed to replace the regular daily losses. Then there is obviously a greater need during growth, when muscles are developing in physical training and when tissue is being restored during convalescence after illness.

To allow for body size the recommended allowance for adults is stated in terms of body weight, 0·57 g of top quality protein per kg body weight. The average 65 kg man (10 stone or 140 lb) would, therefore, need 37 g of protein per day—assuming it has a BV of 100. In Great Britain the average diet has BV of 70 so that the protein needs are 37 multiplied by 100/70 or 53 g.

This is termed the 'safe level' and is the figure adopted by the Committee of the Food & Agriculture Organization in 1973. (Some tables of recommended intakes used in various countries put the

TABLE 12

Protein allowance in terms of reference Protein (F.A.O. 1973)

(proportionately extra is needed for protein of quality score (BV) 80, 70 or 60)

			g protein per kg body weight per day	
Babies age 0–6 months			2·0*	
	6–11 months		1·53	
Children	1 to 3 years		1·19	
	4 to 6 years		1·01	
	7 to 9 years		0·88	
Adolescents				
	10 to 12 years	male	0·81	female 0·76
	13 to 15 years		0·72	0·63
	16 to 19 years		0·60	0·55
Adults			0·57	0·52
Pregnancy plus 9 g per day in second and third trimesters				
Lactation plus 17 g per day				

* As milk protein.

figure a little higher than this to build in an extra safety factor as a matter of national food policy, but we must accept the F.A.O. figure as being adequate.)

In U.K. tables the amount of protein recommended increases with energy expenditure. There is no *need* for more protein for heavier work but there is obviously a need for more *food*. People are recommended to eat more of their usual diet, which is about 10% protein, hence higher energy intake is accompanied by higher protein intake. The figures are therefore more realistic than increasing recommended energy intake while keeping protein figures constant (p. 186).

QUALITY SCORE

Protein quality can be calculated by comparing the amino acids of the diet with a reference pattern. This is termed amino acid or protein score.

WHERE DO WE GET IT?

Food composition tables give the protein content of foods in grammes per oz or per hundred grammes. Values for some of the principal sources of protein are as shown in Fig. 18.

From figures like these it can be calculated that the allowance of 65 g is obtainable from 3 pints (1·75 litres) of milk, or 9 oz (260 g) of beef, or 6 oz (170 g) of cheddar cheese or 9 oz (26 g) of cod. In more practical terms this quantity of protein may be obtained from ½ pint of milk (9 g protein) plus one egg (7 g) plus 2 oz cheese (14 g) plus 6 oz meat (15 g) plus 5 oz bread (11 g).

The same amount of protein, although of poorer quality, may be obtained from 12 oz bread or cereal (18 g protein) plus 10 oz potato (4 g) plus 2 oz pulses (16 g) plus 3 oz peanuts (24 g).

DO WE GET ENOUGH?

The average diet eaten in Great Britain provides nearly 80 g of protein per day—well in excess of the 'safe level'. With the kind of food that is available in Great Britain the protein intake will almost always be adequate so long as the appetite is satisfied. There are two main reasons: first, the staple food, that which makes the major contribution to the diet, is wheat, containing 10% protein. It is rather surprising to find that even on our very varied diet nearly 30% of the energy still comes from bread, flour and cereals. It is probably more

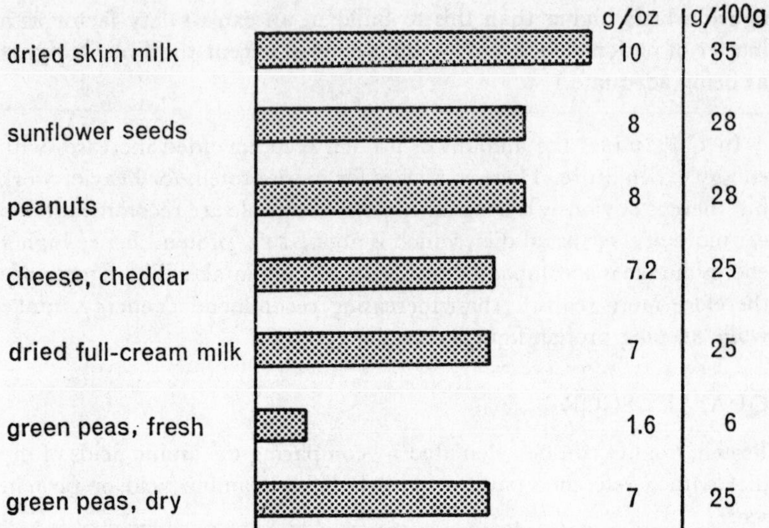
g/oz
g/100g
dried skim milk
10
35
sunflower seeds
8
28
peanuts
8
28
cheese, cheddar
7.2
25
dried full-cream milk
7
25
green peas, fresh
1.6
6
green peas, dry
7
25
haricot beans, dry
7
25
almonds
6
21
beef
5
18
codfish
5
18
evaporated milk
2
7
liquid milk
0.9
3

Fig. 18 Some sources of protein

surprising to some to discover that these foods, although popularly regarded as carbohydrate foods, also supply nearly 30% *of the protein* of the diet. The second reason why our diet is adequate in protein is that we have an abundance of protein-rich foods like meat, milk, cheese, etc.

The position is very different in some of the developing countries where (1) the staple food is cassava, yams, plantains or sweet potato containing only 2 to 3% protein, and (2) very little protein-rich food is available. The problem is discussed in Chapter 9.

SPECIAL NEEDS FOR PROTEIN

The elderly The figure of 80 g per day for protein intake in Great Britain is an average and individuals may, for a variety of reasons, be getting less than this. One such group is the elderly, who, as discussed in Chapter 6, usually eat less than they did when younger and so automatically eat less protein. Moreover, the reduced quantity of food is sometimes composed of the more starchy and sugary foods, so further reducing the protein intake.

If they are short of protein the diets of older people should contain more eggs, milk, fish, etc., and could be enriched with such concentrated protein foods as dried milk, soya, and dried yeast to increase protein value at low cost (see Chapter 6).

Convalescents There are groups of the population who have higher needs for protein, e.g. convalescents.

When the body suffers damage such as broken bones, burns or surgical shock there is an extremely large loss of protein. Not only is there the obvious loss as, for example, in a burn, but the injury causes the body to break down large quantities of protein tissue. This protein is oxidized and lost to the body. The surgical shock involved in removing the appendix can cause a loss of 150 to 230 g of protein, a fractured femur can cause a loss of 800 g of protein, severe burns can result in losses in excess of 1 000 g, and infections can result in similar losses depending on their severity. These losses can amount to a tenth of the total protein content of the body.

Convalescence involves the restoration of this protein. The diet, therefore, must be rich in protein both in quality and quantity. As this need occurs at a time when the appetite of the patient may be poor, convalescence poses a serious nutritional problem. Special protein-rich foods such as milk casein, which is 95% protein, and certain proprietary preparations, are valuable additions to the diet. At the same time, care must be taken to ensure that enough energy-giving foods are also provided, otherwise part of the valuable protein may be 'wasted' by being oxidized to provide the energy that the body must always have.

Babies Another group of the population with special protein needs are babies at weaning. During the rapid stages of growth, protein needs are relatively high but are adequately provided by milk, either from the breast or the bottle. When the baby is weaned on to solid foods care has to be taken that there is enough protein in the diet; and milk still remains the basis of his diet. When milk is not freely available and weaning foods are inadequate in protein content, the baby's growth may be affected. The problem is one of major importance in the developing countries and is discussed in Chapter 9.

PLANNING AROUND PROTEIN

The feature of protein foods as the centre of a meal has long been recognized—even if not from the nutritional point of view. Meals are planned round the main protein foods and large intakes of meat, fish and eggs are often associated with wealth.

When meals are prepared for large numbers of people cost often limits the amount of protein and cheap meals are apt to contain over-large amounts of carbohydrate. To avoid this school meals are designed to supply between one third and one half of a child's daily recommended intake.

The choice of foods for construction and repair can be made expensively by using meat, fish, cheese, milk and eggs, more cheaply by using cheaper cuts of meat and cheap fish (which have the same nutritive value as the more expensive varieties) and more cheaply still with vegetable foods. There are many traditional dishes which illustrate the use of small amounts of the more expensive and protein-rich animal foods as a supplement to cereals; e.g. a large amount of rice supplemented with a small amount of meat, milk, cheese or fish as in a risotto, pilaff, paella, kedgeree and rice pudding. Similarly good dishes are made with pasta such as macaroni-cheese and spaghetti bolognaise. Meat pasties, puddings and pies are further examples.

In some countries the traditional way to plan a meal is to have a large amount of cereal or root vegetables and to serve small dishes of spiced meat or fish and a sauce, such as soy sauce, or namplu (made from shellfish) which provides the protein. This can be a good pattern nutritionally if the relishes are well chosen and adequate in quantity.

Dried skim milk is an excellent supplement for a low-protein diet. It can be used in liquid form, or mixed dry into doughs and batters, vegetable soups and stews.

'BEST BUYS' FOR PROTEIN FOODS

Caterers are often allowed only a limited sum of money per head to spend on meat, fish, etc. The amount of protein provided per penny is a useful guide in spending this amount to advantage (see Table 13 and p. 187).

Such a list, kept up to date, will help caterers in buying and planning economical meals of good protein value.

When the cheapest sources are found then recipes must also be found to put them over attractively, e.g. coley plus tomato, in fish cakes or pie; minced meat as hamburgers and kofta (meat ball) curry; breast of lamb, steamed, boned and made into pasties; offals, e.g. liver, heart, melts, and kidney, made into faggots (see Appendix p. 199).

TABLE 13
Cost of protein foods

Food	*Cost per lb* *pence*	*Waste* %	*Protein* *g per oz*	*Protein* *g per penny*
Breast of lamb	25	50	4·5	1·4
Beef sausages	45	—	2·7	1·0
Ox liver	28	—	5·9	3·3
Heart	45	10	4·9	1·5
Brisket, cheap end	55	25	4·8	1·1
Streaky bacon, cheap	75	12	4·1	0·8
Minced beef	60	—	5·8	1·5
Coley fillet	60	10	4·5	1·1
Cheese	58	—	7·2	2·0
Dried skim milk	56	—	10·2	2·9
Chicken	38	31	5·8	1·7

Large cheap joints of meat can be made tender and tasty by long, slow cooking in an oven at 275 °F (135 °C), allowing approximately 1 hour per pound of meat (estimated weight without bone). This method not only produces attractive meat with less shrinkage but also a clean fat which can be used for making pastry. The meat is cooked when an oven thermometer registers 170 °F (77 °C) in the thickest part.

To get the greatest protein value from these foods they should be served with cereal such as flour or rice, and vegetables such as potatoes, which will supply the necessary energy. Such mixtures will also have an enhanced protein value through complementation between the different proteins.

VEGETABLE PROTEINS

In parts of the world where animal foods are not available, protein must come from vegetable sources. Peas, beans and many nuts are relatively concentrated sources of protein and can adequately supplement a cereal diet. Where the staple food is one of the protein-poor foods such as starchy roots, then it is difficult to ensure an adequate intake of protein unless care is taken to select the concentrated sources.

Some typical vegetable mixtures of good protein value are Kitchri (lentils steamed with rice and butter); West Indian dishes of rice and peas; *dahl-pouri* (pancakes of flour with filling of dried pulse); *milk loaf* using flour, yeast and two ounces of dried skim milk per pound of flour (150 g per kg). Another way of providing more protein is from textured vegetable protein, which includes meat analogues made from soya. The mixture is extruded or spun to give a texture like meat. It is flavoured, mixed with fat and coloured. These products lack the B-vitamins and iron found in meat, but these nutrients can be added, for example with curry, wholegrain cereals and pulses to make appetizing food. See p. 198.

PRACTICAL WORK

1. Calculate the weights of the following foods needed to provide 10 g of protein: fresh milk, evaporated milk, gammon bacon, soft roes.
2. Make a block diagram of the amounts of protein supplied by the following amounts of food: bread (10 oz—280 g), potatoes (10 oz—280 g), lean beef (3 oz—85 g), lean bacon (2 oz—57 g), cheese (1 oz—28 g), milk (½ pint—285 ml).
3. Make up a drink containing 10 g protein.
4. Prepare a list of meals which would supply the bare minimum of protein needs for the day, then improve the recipes by adding protein supplements and calculate the improved content.
5. Compile a recipe of ingredients to make up a protein-rich powder which could be used in an emergency, i.e. mountaineering, skiing, etc. Make practical suggestions how the powder could be used.
6. Make up a protein-enriched batter mix and give directions for use.
7. Suggest ways of improving the protein value of the following recipe. Dried potatoes or yams, carrots, swedes, with beans, cooked in water with onions and herbs.

EXAMINATION QUESTIONS

1. Discuss the importance of proteins in the diet. What are their main sources? Describe the processes involved in their digestion. (City & Guilds Adv. Dom. Cookery Theory.)
2. Discuss the cost of different sources of protein in relation to their nutritional value. (R.S.H. Certificate.)
3. A catering office consults you about 'Body-Building Foods' for the dietary of a children's ward. What advice would you give? (R.S.H. Certificate.)
4. Is it necessary to include some animal protein in an adult's diet, and if so, why? State the amount which it is usual in this country to include and name the foods and quantities which will supply it. (R.S.H. Certificate.)

5. State the functions of protein in the body. Plan one day's meals for a male clerk, giving the weights of portions served, to show how you would meet his protein requirements for the day. (R.S.H. Certificate.)
6. 'Protein foods are important in the diet, but usually expensive.' Discuss this statement. What advice on food choice would you give a young housewife on a limited budget? ('O' level, London.)
7. Explain how you would give an adequate amount of protein to a retired man of 65 to 70 years, for 2 days. Discuss the value of the proteins from the different sources you have used in relation to cost. (Hotel & Catering Inst., Inter.)
8. What in your opinion are the cheaper ways of providing good quality protein? Plan 3 days' meals for an old age pensioner, which would be within his purchasing power, and would be nutritionally adequate. Calculate the protein, calories and vitamin C for one day. (R.S.H.)
9. Lysine is the limiting amino acid in cereal proteins e.g. rice, wheat, maize and rye. Some good sources of lysine are animal protein foods, pulses (or legumes) and soya flour. Illustrate how these facts can be put to practical use by preparing 2 dishes of good protein value for teenage boys and 2 dishes for vegetarian adults. Prepare and serve a three-course lunch for vegetarian office workers using some of the dishes. Calculate the energy and protein value of one of the dishes. ('A' level practical Home Economics, Cambridge.)
10. Compare the quality of animal and vegetable proteins, and comment on the nutritional aspects of 2 advertising campaigns: 'add an egg' and 'mix with milk'. (R.S.H. Certificate).

Part 2 CALCIUM

BONES

The body is constructed of a framework of bone, and physical strength and a good figure depend to a large extent on a well-built skeleton and strong bones. Poor bone formation can affect the pelvic girdle and cause difficulties during pregnancy and childbirth.

Three nutrients are needed to form bone: calcium and phosphate as the principal structural materials and vitamin D to assist the absorption of the calcium from the digestive tract into the blood stream. Bone is based on cartilage, which is a protein, on which is deposited a mixture of calcium phosphate and calcium carbonate. Mature bone is about 50% mineral salts, 20% protein and 25% water.

These two parts of bone can be shown by a simple experiment. When a bone, such as chicken leg bone, is burned over a clear flame the protein is burned away and the calcium phosphate is left as a white ash. If instead the bone is treated with dilute acetic acid or vinegar for 2 to 3 hours, the mineral matter is dissolved away leaving the cartilage looking like a jelly-bone.

Bone growth The bones of babies and young animals are largely cartilage. The bones of children continue to lengthen and harden up to the age of about 19 years when they amount to one-sixth of the total body weight, 10 kg. Since the newborn baby's bones contained only about 40 g of calcium it is clear that a large amount must be supplied in the diet.

As well as the calcium and the phosphate used as structural materials proper bone formation needs vitamin D to ensure the absorption of the calcium and to aid its deposition on the cartilage, and vitamins A and C play a part in the growth of the cartilage.

These nutrients are still required by the mature adult as, even after the bones have been formed, there is a continuous removal and renewal of calcium phosphate. This has been demonstrated by feeding foods 'labelled' with radioactive calcium.

During pregnancy and lactation there is a greatly enhanced need for calcium to form the bones of the new baby. If the diet of the mother is poor then calcium is withdrawn from her bones to supply the needs of the baby. Repeated pregnancies and poor diet can severely weaken the bones and teeth of the mother.

TEETH

Teeth are similar in composition to bone and require the same nutrients. In addition to the calcium, phosphate and vitamins, the mineral element fluorine helps to make the strong protective enamel. This is often present in drinking water in adequate amounts, about one part per million; but if it is not then the enamel is weak and the teeth soon decay. This is the reason for adding fluorine (in the form of fluoride) to drinking water in many parts of the world. Where this has been done the rate of decay of children's teeth has been reduced to about half.

SOURCES OF CALCIUM

All diets appear to contain enough phosphate so this mineral does not need to be considered further. Calcium, however, poses a real problem as the only rich sources are milk and cheese, and fish that is eaten with its bones, sprats, whitebait, dried and canned fish. Calcium is added to some foods during preparation, such as tripe and molasses, and one of the main sources in our diet is white bread which is enriched by the addition of chalk (calcium carbonate). (See pp. 156 and 159).

Absorption of calcium A further problem associated with calcium is that it is rather poorly absorbed; only 20 to 50% of the calcium that is consumed is absorbed from the food and the rest passes right through the intestine and is lost to the body. Several factors can affect this figure, the most important of which is vitamin D. Its absence gives rise to rickets in children because not enough calcium is absorbed for proper bone formation. But even when vitamin D is present there are other interfering substances in the diet. One of these of minor importance is oxalic acid. This can combine with calcium to form insoluble calcium oxalate which is not absorbed. It is found in foods such as spinach, chard, sorrel, green bananas and rhubarb, and calcium oxalate can sometimes be seen as a white precipitate in jars of bottled rhubarb. Oxalate is, however, usually of minor importance.

A factor of much greater importance is phytic acid. This is found in the branny part of cereals and, therefore, in brown flour (but not in white flour) and also in nuts and pulses. It is not clear just how effective phytic acid is in preventing the absorption of calcium as it is counteracted to some extent by the enzymes of yeast. So when brown flour is made into a dough with yeast some of the phytic acid is destroyed. It is similarly counteracted by the enzyme phytase which develops in pulses when germination starts during soaking in water. So the effect of phytic acid can be reduced by cooking with yeast, by using a long cool rise for wholemeal breads, doughs and batters, and by steeping pulses before use.

Absorption of calcium is also reduced by the laxatives *Glauber's salts* and *Epsom salts* since these form insoluble calcium sulphate.

Some dietary factors assist the absorption of calcium from our food, such as protein and the organic acids from fruits and vegetables. Therefore, milk and cheese are exceptionally useful since they contain both calcium and protein. Other useful combinations are fruit 'fools', vegetable soups made with milk and white sauces served with vegetables, cheese and pickles.

HOW MUCH CALCIUM?

To take care of the relatively poor absorption of calcium an intake of 500 mg a day is recommended for adults. This is the figure suggested by the Expert Committee of the Food & Agriculture Organization

TABLE 14
Calcium content of foods

Food	*Calcium*	
	mg per oz	*mg per 100 g*
Milk products (cow)		
Dried, skim	360	1 260
Dried, whole	280	1 020
Fresh	35	120
Cheese, Parmesan	350	1 220
Cheddar	230	800
Edam/processed	210	740
Stilton	100	350
Buffalo milk, fresh	60	200
Yogurt (average)	50	180
Fish		
Dried, with bones	850	3 000
Whitebait, fried	240	860
Sprats, whole, fried	170	600
Sardines	150	550
Pilchards, canned	85	300
Various products		
River snails	430	1 500
Sesame seeds	430	1 500
Finger millet	100	350
Soya beans	60	200
Tripe, raw	20	75
Molasses	140	500
Almonds	70	250
Figs, dried	75	280
Kale	70	250
Broccoli	30	100
Haricot beans	50	180
Enriched white bread	30	100
Enriched white flour	40	150
Unenriched white flour	5	15

(Report No. 30) and agreed in the U.K. Tables and is rather less than the figure given in earlier food tables (800 mg).

The rapid bone growth of children calls for a higher calcium intake and the figure is 600 mg for babies of 1 year and less, 500 mg for children of 1 to 9 years, increasing to 700 mg for children of 10 to 15 years. The figure falls to 600 mg for the 16 to 19 age group and finally to the adult figure of 500 mg per day.

In pregnancy and lactation a much higher intake of 1 200 mg is recommended.

SOURCES OF CALCIUM

A food can be considered a good source of calcium if a portion supplies one-fifth of a day's allowance or more, i.e. 100 mg.

The easiest way to ensure an adequate supply of calcium in the diet without checking every meal is to arrange for one calcium-rich food to be eaten each day. Calcium intake can also be increased by improved methods of cooking: e.g., by preparing fish and meat with acids such as vinegar and lemon juice which dissolve the calcium from the bone (sweet-sour spare-ribs of pork; West Indian souse, in which pigs' trotters are cooked, and then steeped in vinegar and lemon juice with herbs and served with cucumber and lemon; potted and soused fish).

As a public health measure in several countries, calcium is added to wheat flour with the result that bread becomes one of the major sources of this nutrient.

PRACTICAL WORK

1. Weigh out food portions providing 200 mg of calcium.
2. Cost these portions; which are the cheapest?
3. Compare the amount of calcium obtained per unit spent: (a) sardines at 23p per $3\frac{1}{2}$ oz tin (100 mg) and (b) pilchards at 28p per 10 oz (280 g) tin. Which is the best buy?
4. Suggest two attractive ways of using the sardines and pilchards.
5. Make a block diagram of the calcium allowances of the various age groups.
6. Make up a bread recipe of improved calcium value, and calculate the amount of calcium per slice, and the fraction of the day's allowance in 4 oz (112 g).
7. Collect recipes providing approximately 200 mg calcium per portion.
8. Plan a meal providing 500 mg of calcium.
9. Plan a meal low in calcium.
10. If someone does not like fish, milk or cheese, how would you provide enough calcium in the day's meals?
11. Collect recipes making good use of dried skimmed milk.

Part 3 VITAMIN D

To construct strong teeth and bones it is necessary to have enough calcium *and* enough vitamin D for its absorption. Even if there is sufficient calcium in the diet not enough will be absorbed if vitamin D is lacking.

The vitamin can be obtained from two entirely different sources; it can be supplied in the food and it can be made in the skin under the influence of ultraviolet light. The natural form that is found in foods and is formed under the skin is called vitamin D_3 or cholecalciferol. The ultraviolet light from sunshine and from lamps is equally effective. Vita glass allows these rays through but they are reduced by ordinary glass, smoke and mist. This is the reason why rickets, the disease that results from vitamin D deficiency, was common in northern climates where little sunshine gets through: it is less common now that vitamin D is added to foods and smoke abatement is enforced.

The synthetic form, which is just as useful to us as the natural form, is vitamin D_2 or ergocalciferol. It is manufactured by ultraviolet irradiation of the substance ergosterol which is extracted from yeast. This is the form of the vitamin that is usually added to foods such as dried and evaporated milk, baby foods and margarine. (There is no vitamin D_1, the name was originally given to an impure mixture.)

RICKETS

Rickets is the result of a shortage of vitamin D together with a poor diet so that there is not enough calcium present in the blood stream for calcification of the bones. It used to be common in Great Britain and in the past centuries was known as the 'English disease'.

Improper calcification of the bones means that they remain soft and when the baby begins to walk they bend under his weight. Knock-knees and bandy legs are typical of severe cases. Other signs of rickets are a 'bossing' of the forehead and 'pigeon' chest.

Early signs of rickets can be seen by X-ray of the bones, as they are less dense than normal bones, or by a biochemical test for the enzyme alkaline phosphatase. This is normally present in the blood stream but its level is increased threefold in rickets.

Since vitamin D has been generally available, originally as cod liver oil and later in various proprietary preparations, rickets has largely disappeared from Great Britain. A few cases have occurred recently because of improper feeding of the babies. The mothers had used the ordinary liquid milk which is not fortified with vitamin D and failed to give the baby cod liver oil or a similar preparation.

Another and more severe problem is that of rickets among Asian immigrants into Britain. This particularly affects children and adolescents of both sexes and also pregnant women.

It is not clear why these individuals are short of vitamin D—it may be a dietary shortage, lack of exposure to sunshine, something in their diet that interferes with calcium metabolism, or even genetically high requirements. The problem is very complicated because firstly, it is found mainly among Asians, but not West Indians, in Britain, and secondly it is relatively common in many sunny countries like Iran and the Indian subcontinent. It is not clear, though, whether people are actually exposed sufficiently to the sun in such sunny countries for this to make any difference.

Meanwhile it is successfully treated by extra vitamin D in the diet.

FUNCTION OF VITAMIN D

Cholecalciferol from the diet or from the skin passes to the liver where an enzyme inserts a hydroxy group (OH) on carbon number 25 to form 25-hydroxy cholecalciferol. This then travels to the kidney where a second hydroxy group is inserted, this time on carbon number 1 to form 1.25 dihydroxy cholecalciferol. This compound is the active form of the vitamin and stimulates the absorption of calcium from the food in the intestine into the blood-stream. It also has some control over the laying down of calcium on bone.

This recently established biochemistry of vitamin D explains an old problem, that of vitamin-D-resistant rickets. As the name implies, this is a form of rickets that fails to respond to vitamin D. It is now known that it is due to the hereditary absence of the enzyme that converts vitamin D at the first stage into 25 hydroxy CC. Such patients can now be cured by giving them this compound.

It also explains why patients with chronic kidney disease develop bone disease.

OSTEOMALACIA

A disease equivalent to rickets in adults, known as osteomalacia, can sometimes occur in rather special circumstances. Women living on a poor diet, low in calcium and vitamin D, and not exposed to sunshine, fall victims when they have repeated pregnancies. The formation of the baby's bones takes priority and calcium is taken from the mother's bones. Osteomalacia occurs particularly in tropical countries among middle-class Muslim women who live in 'purdah', i.e. indoors, and when they do go out they are heavily veiled and protected from the sun. It occurs also in Great Britain in elderly housebound women living on a vitamin D-poor diet.

OSTEOPOROSIS

Osteoporosis differs from osteomalacia since there is a diminished amount of bone, which is, however, still fully mineralized. It is common in women over about 50 and leads to weak bones which are easily fractured.

It used to be considered to be a hormonal rather than a dietary problem, but recent work suggests that it may be related to diet or at least alleviated with large amounts, about 1.5 g per day, of calcium and vitamin D. It is considered by some to be related to osteomalacia.

HOW MUCH VITAMIN D?

Vitamin D is measured in micrograms (although older textbooks use international units—1 microgram is equivalent to 40 i.u.).

Since vitamin D can be obtained either from the diet or from the effect of sunlight on the skin the requirements are not clearly established. It is not possible to measure how much is made in the skin and the amount will certainly vary with the amount of clothing and social conditions as well as the amount of sunlight available. Children living in the dark, narrow slum streets of cities like Calcutta develop rickets, although the majority of the people in the tropics get enough vitamin D from the sunlight on their skin.

The recommended daily allowances are:

Infants	10 μg	(400 i.u.)
Children and teenagers	2·5 μg	(100 i.u.)
Adults	2·5 μg	(100 i.u.)
Expectant mothers	10 μg	(400 i.u.)
Nursing mothers	10 μg	(400 i.u.)

SOURCES

Table 15 shows that only fatty fish and eggs serve as a reasonable source of vitamin D among the unenriched foods. To remedy this state of affairs margarine is enriched with vitamin D by law in many countries; and evaporated milk and baby foods are enriched voluntarily.

Vitamin D is stored in the liver and menus can be planned to provide about 20 μg per week rather than on a daily basis, using figures in Table 15.

Since the fragile bones of elderly people improve to some extent on a diet rich in calcium and vitamin D their diet should include adequate amounts of these nutrients. The cheaper fatty fish could be popularized, and cod liver oil and halibut oil can be mixed into dishes such as fish

TABLE 15
Sources of vitamin D
(weighed without waste)

	μg *per oz*	μg *per* 100 *g*	*i.u. per* 100 *g*
Cod liver oil B.P.	70	250	10 000
Eels, raw	36	125	5 000
Brisling	14	50	2 000
Herring, kippers, etc., raw	6	22	900
Mackerel, raw	5	18	700
Salmon, canned	3·5	12·5	500
Sardines	2·5	9	320
Tuna, canned	1·7	5·8	230
Egg, whole	0·5	1·7	70
Egg yolk	1·4	5·0	200
Cod liver oil and malt	6	22	880
Margarine	2·2	8	90
Butter	0·2	0·7	35
Liver, ox	0·3	1	40
Cod's roe, hard	0·5	2·0	70

Portion of food supplying 20 μg, i.e. 1 week's allowance

Brisling	1½ oz (5 small fish) (42 g)
Herring fillet	3 oz (2 fillets) (85 g)
Kipper	3 oz (2 fillets) (85 g)
Mackerel	4 oz (114 g) (flesh of one fish (7 oz, 200 g))
Eels, raw	½ oz (14 g)
Cod liver oil	2 teaspoons
Salmon	5½ oz (1 slice or 1 small tin) (155 g)
Egg yolks	4 oz (2 eggs) (115 g)
Sardines	7 oz (1 large tin) (200 g)
Margarine	8 oz (230 g)
Cod liver oil and malt	3 oz (2 tablespoons) (85 g)

pies. Useful amounts of fish liver oils can be mixed into bread, bun doughs and ginger bread mixtures either as the oil or as malt-and-cod liver oil preparations. One teaspoon of the oil or a tablespoon of the malt and oil per lb of flour has little or no effect on the taste.

STABILITY

Vitamin D is stable to heat and storage and little is lost in most methods of food preparation. Cod liver oil goes rancid when exposed to air, warmth and light and so should be protected from these by storage in a closed, dark container in a cold place.

CAN WE EAT TOO MUCH CALCIUM OR VITAMIN D?

We can scarcely have too much calcium in the diet since the intestine will exercise some control over the amount absorbed. In the case of vitamin D it is possible to obtain too much, since many foods, particularly baby foods, are enriched with added vitamin D. In the past a small number of babies have been found to be suffering from excessive amounts of calcium in the body, believed to be caused by excessive dosing with vitamin D. The disorder is called hypercalcaemia. They were being given 75–100 μg per day from fortified milk, cereals, cod liver oil and baby foods. Such quantities are well tolerated by the vast majority of babies but these few were hypersensitive. To prevent hypercalcaemia mothers who give their babies D-enriched foods are recommended to give them only half instead of a whole teaspoon of cod liver oil.

PRACTICAL WORK

1. Weigh out portions of food providing 20 μg of vitamin D.
2. Calculate their cost. Which is the cheapest source?
3. Calculate the amounts of vitamin D and calcium per portion in three fish menus. What proportion of the daily needs of young people do they supply.
4. Suggest a diet for an expectant mother which will contain enough calcium and vitamin D without the use of medicinal or vitamin preparations.
5. What difference would it make to the vitamin D content of the average British diet if margarine were not enriched with this vitamin?
6. Compile a leaflet to persuade and help an Asian or West Indian mother, living in the north of England to give her children enough vitamin D.
7. How would you make a significant increase in the vitamin D and calcium content of these dishes:
   (a) Baked casserole of white fish, with tomatoes?
   (b) Gingerbread?
   (c) Mixed salad of cucumber, peppers, tomatoes, olives with oil and vinegar dressing?

EXAMINATION QUESTIONS

1. What are the main sources of calcium in the diets of (a) infants, and (b) adults? What factors influence the absorption of calcium

from the intestine? How is the requirement for calcium influenced by (a) growth, (b) pregnancy, and (c) lactation?

2. Discuss the principal dietary factors and/or foods which: (a) are concerned in the formation of good teeth, and (b) are responsible for adequate blood formation.
3. What nutrients are needed in the diet to build and maintain healthy bones and teeth? How would you provide these nutrients when planning menus for young children? (R.S.H.)
4. Rickets was so common in England in the Middle Ages, that it was known as 'The English Disease'. Explain why rickets has virtually disappeared today. (R.S.H.)
5. Which foods provide the best sources of (a) vitamin D, (b) calcium, and (c) iron? Why is it necessary to have regular supplies of these and what is the result if they are lacking? (City & Guilds 244, Adv. Dom. Cookery.)

SUGGESTED MENUS TO SUPPLY CALCIUM, VITAMIN D AND PROTEIN

1. *Fish dish hot*

Grilled mackerel or herring, roes fried in batter, fish cake, served with chips, watercress, tomatoes. Biscuits and cheese and coffee.

2. *Fish dish cold*

Salad plate including salmon mayonnaise, rollmop, sardines, cucumber, lemon, watercress.

3. *Seafood plate*

Stuffed pancake with filling of smoked salmon, hardboiled egg, mussels, etc. Bechamel sauce with grated cheese topping.

4. *Methods of serving cheaper fatty fish*

To avoid the objections of smell and bones the fish is filleted and the small bones softened by marinading in vinegar or lemon juice, or cooking with tomatoes.

Alternatively, the raw fish is minced, using a mixture of fat fish and white fish, and mixed with egg yolk, butter, potato or bread crumbs to make fish cakes, or 'seaburgers' fried like hamburgers.

If mackerel and herring are cooked with herbs and onions the fishy smell is masked by the savoury flavour.

Herrings can be served in a variety of ways as *hors-d'œuvres*, bucklings, kippers, rollmops, and pâtes.

Part 4 VITAMIN A

Vitamin A exists in several forms: carotene and related pigments which are found in vegetables and fruits, and retinol in animal foods. Earlier the animal form was termed vitamin A and carotene was termed a pro-vitamin, since it is converted into retinol in the body. In modern nomenclature each substance is given a specific chemical name and vitamin A refers to all substances that have the particular biological action.

Vitamin A was the first vitamin to be named (in 1915) when it was found that a substance in milk, butter, and, later, watercress was essential for growth of animals. It has several functions and the earliest sign of a shortage is a difficulty in seeing in dim light, called night blindness.

In dim light, as distinct from bright light, we see with the aid of a pigment in the cells of the retina called visual purple or rhodopsin. This is a protein linked with retinol. After we have been in bright light it takes a little time to get used to the dark—this is because the bright light bleaches the visual purple and it takes a few minutes to be remade. When vitamin A is in short supply it takes much longer or it may never be completely remade.

The time taken to recover in this way is the basis of the 'dark adaptation test' used to detect early stages of vitamin A deficiency.

A great deal of retinol—the amount depends on the diet—is stored in the liver, and the blood level is kept up so long as there is some in store (in the same way as there is water in the pipes so long as there is some in the storage tank). So there are no outward signs of vitamin A deficiency even on a poor diet until these stores are nearing exhaustion.

Night blindness is quickly reversed by feeding retinol or carotene but some of the later changes are not. The tear glands of the eye become blocked, and the conjunctiva, the moist protective covering, becomes dry and inflamed. This condition is called xerophthalmia (literally 'dry eye'), and is reversible.

Later, the cornea, the thick transparent skin covering the eyeball, becomes ulcerated. This is keratomalacia and is not reversible, and leads to blindness. Vitamin A deficiency is a common cause of blindness among children in many developing countries where they are fed on cereals, sugar and *skimmed* milk—all of which are devoid of vitamin A.

In addition to its function in the eye and in growth, vitamin A is concerned with the formation of the viscous fluid (mucopolysaccharides) that keep the tissues lining the nose, throat, tear ducts and other ducts of the body in their moist condition.

Animal foods are listed in food composition tables as containing so many i.u. or μg retinol per oz or per 100 g. Vegetable foods may be shown as containing so many i.u. or μg carotene, or may have been converted into i.u. vitamin A (older textbooks) or μg Retinol Equivalents. Some foods contain both retinol and carotene, and calculation of the vitamin A potency in terms of Retinol Equivalents depends on the way the figures are provided. *We need to convert ALL figures into* μg R.E.

Aid to Calculation and Conversion into μg R.E.

(1) μg beta-carotene into μg R.E. . . . *Divide by 6.*
(2) i.u. beta-carotene into μg R.E. . . . *Divide by 10.*
(3) i.u. retinol into μg R.E. . . . *Divide by 3·33 or Multiply by 0·3.*

For example, egg contains both carotene (30%) and retinol (70%). Figures may be provided in three ways: (A) μg of each; (B) i.u. of each; (C) i.u. of total vitamin A activity where carotene figures have already been converted into retinol.
We need to convert all figures into μg R.E.

Example A	μg	μg R.E.	
Retinol	210	210	= 300 μg R.E.
Carotene	540	540 ÷ 6 = 90	

Example B	i.u.	μg R.E.	
Retinol	700	700 × 0·3 = 210	= 300 μg R.E.
Carotene	900	900 ÷ 10 = 90	

Example C	i.u. vitamin A activity		μg R.E.
Retinol	700	1000	1000 × 0·3 = 300 μg R.E.
Carotene	300		

Recommended intake

		μg R.E.	*i.u.* (vitamin A)
Babies	**6 to 12 months**	**300**	**1 000**
Children	**1 to 3 years**	**250**	**830**
	4 to 6	**300**	**1 000**
	7 to 9	**400**	**1 330**
	10 to 12	**575**	**1 920**
	13 to 15	**725**	**2 420**
Adults and children over 16		**750**	**2 500**
Pregnancy		**no increment**	
Lactation		**1 200**	**4 000**

In the U.K. mixed diet we consume about one-third as retinol and two-thirds as carotene.

SOURCES

Carotene, deep orange in colour, occurs in carrots, coloured fruits apricots, peaches, pawpaw and melon. All the green parts of pl contain carotene, although the colour is masked by the green chlorophyll. Red palm oil is an extremely rich source of another fo of carotene called alpha-carotene, and yellow maize contains another form called cryptoxanthin. Both these forms of vitamin have only half the biological activity of the common form of caroter beta-carotene.

Retinol occurs only in animal foods like milk, butter, eggs and liv (together with some carotene). The animals (and man) obtain thei retinol either from animal foods or by conversion of carotene. Cod live oil and halibut liver oil are rich sources of retinol because the fish feec on algae, which contain carotene. Margarine has retinol added in many countries.

MEASUREMENTS AND REQUIREMENTS

Vitamin A used to be measured in international units (i.u.). It is now measured in micrograms (μg).

The amount of carotene that is absorbed and then converted into retinol in the wall of the intestine varies with the food (e.g. better absorbed from puréed carrot than whole carrots) and also with the different chemical forms of carotene. So vitamin A is measured in foods in retinol equivalents (R.E.), that is the carotene or other form of vitamin A is multiplied by a factor to allow for incomplete absorption and conversion into retinol.

Definitions:

1 μg retinol equivalent (R.E.) is equivalent to:

(1) 1 μg retinol.
(2) 3·33 i.u. retinol (in old nomenclature 3·33 i.u. vitamin A).
(3) 10 i.u. beta-carotene.
(4) 6 μg beta-carotene.

These carotene conversion factors are used for all foods with two exceptions:

(1) beta-carotene is so well absorbed from milk that 2 μg carotene = 1 μg R.E.
(2) The vitamin in red palm oil (which is alpha-carotene), and that in yellow maize (which is cryptoxanthin) are only half as active biologically as beta-carotene. Therefore 12 μg or 20 i.u. of vitamin A from these sources = 1 R.E.

DO WE GET ENOUGH?

On the sort of diet that we get in the U.K. there is no fear of a shortage of vitamin A since almost everyone has milk, butter, margarine, cheese and green vegetables and carrots. Since any surplus of vitamin A is stored in the liver and since both carrots and liver are so rich a source either of these foods eaten even only once a week would supply all our needs of the vitamin. So it is most unlikely that we could go short.

In some countries where dairy foods are not available, meat supplies restricted and liver is the butchers' perquisite, vitamin A is hard to come by. Red palm oil, where it is available, is an extremely rich source of carotene, otherwise the population must rely mostly on vegetable and fruit sources (see Table 16).

TABLE 16
Sources of vitamin A (raw, E.P.)
(μg retinol equivalents per 100g)

Retinol		*Carotene*	
Halibut liver oil	600 000	Red palm oil	6 000
Cod liver oil (BP)	18 000	Carrots, old	2 000
Sheep liver	18 110	new, canned	1 000
Ox liver	17 000	Spinach	1 000
Calf liver	14 000	Sweet potatoes	670
Butter, summer	1 300	Apricots, dried	600
average	1 000	Watercress	1 500
Margarine	900	Broccoli	420
Egg yolk	400	Tomatoes	100
Cream,* summer	200	Plantains	10
winter	150	Maize, sweetcorn	30
Cheese, cheddar	350	Mango	200
Edam (low fat)	240	Papaya, canned	80
Stilton	408	Sweet melon	330
Eggs, whole	140	Red peppers	250
Milk. Jersey, summer	50	Cabbage	50
winter	35	Cape gooseberry	200–
Fresian, summer	40		1 200
winter	30	Dried chillies	60–
evap./condensed	100		6 000
* Single		Green peppers	40
		Lettuce	170

ABSORPTION

Both retinol and carotene can be absorbed only in the presence of fat, and there are some parts of the world where vitamin A deficiency is caused not so much by a shortage of the vitamin as by a shortage of fat in the diet. Such dietary conditions are almost impossible in Great Britain but the same principle applies: for example, carotene is better absorbed, about 75%, from margarine, than it is from cabbage, about 25%.

Some methods of food preparation increase the absorption, e.g. pounding, mincing, liquidizing, and the addition of fats such as when broccoli and kale are served with butter or when vegetables are cooked in fats, Chinese fashion, and salads served with an oil, mayonnaise, etc.

Absorption is decreased in liver disorders and any condition that reduces fat digestion. It is also decreased by taking liquid medicinal paraffin at the same time, since the fat-soluble vitamins dissolve in the mineral oil and pass out of the body—mineral oil is not digested.

LOSSES ON COOKING AND STORAGE

Both carotene and retinol are stable to most methods of cooking although there is some loss on frying. They are dissolved in the fatty part of the foods and if the fat is allowed to become rancid by exposure to light, warmth and air, there will be a loss of vitamin A. This is why some brands of butter and margarine are wrapped in foil. For the same reason, fish-liver oils and other preparations of fat-soluble vitamins should be stored in dark glass bottles in the cold.

Vegetables lose a large part of their carotene when they are air-dried as in the old-fashioned methods, but much less when they are dried rapidly at a low temperature as in modern factory processing.

PROVIDING VITAMIN A

Since vitamin A is stored in the body there is no need to plan daily supplies; it is much simpler to plan on a weekly basis. About 5 mg is needed and can be provided by the following:

1. $3\frac{1}{2}$ pints (2 l) of winter milk, 14 oz (400 g) margarine—4·2 mg (14 000 i.u.).
2. 2 oz (57 g) liver *or* 2/3 oz (20 g) cod liver oil—4·2 mg (14 000 i.u.).
3. 14 oz (400 g) carrots, 14 oz (400 g) spinach, 1 lb (450 g) tomatoes, 4 eggs (7 oz), 14 oz (400 g) cheese—3–4·5 mg (10 000–15 000 i.u.).

OTHER VITAMINS

The vitamins discussed here, A, B_1, B_2, niacin, C, and D, are those of nutritional importance since dietary deficiencies do occur. There are two more vitamins that might also be included in this group because there is some evidence of a marginal deficiency, especially during pregnancy, namely vitamin B_6 (pyridoxine) and folic acid.

Pyridoxine is necessary in protein synthesis and also in the formation of niacin from tryptophan. It is found in cereal, eggs, meats (especially liver and kidney), fish, and certain vegetables.

Folic acid is necessary for the formation of desoxyribonucleic acid (DNA) and thus for cell nuclei; deficiency results in a form of anaemia. It is exceptionally rich in liver and also occurs in vegetables, cereal grain, and nuts.

The other vitamins are of medical and biochemical rather than nutritional importance since dietary shortages are very rare. These include vitamin B_{12} (cyanocobalamin), vitamin E (tocopherol), vitamin K (phylloquinone), vitamin H (biotin), and pantothenic acid.

PRACTICAL WORK AND QUESTIONS

1. Prepare two salad plates each supplying approximately 300 μg R.E. (3 000 i.u.) of carotene and including lettuce. Display each salad plate with a card stating the weight of vegetables used and the amount of carotene supplied. Suggest a suitable dressing and give reasons for your choice.
2. Give recipes for and prepare two cream soups each supplying at least 100 μg R.E. (1 000 i.u.) of carotene, 150 μg R.E. (500 i.u.) retinol per portion.
3. How would you improve the vitamin A value for this meal: fried fish and chips, peas, pineapple jelly and banana?
4. Why are condensed and dried skimmed milks labelled 'unfit for babies'? Can you suggest ways by which they could be made suitable for infant feeding.
5. Plan two groups of foods which could supply enough vitamin A for a week for a youth of 18 years. Compare the cost and food value of the two plans.
6. Since liver is such a good source of vitamin A, prepare three recipes using it in an attractive way. Explain how your recipes ensure maximum absorption of the vitamin.
7. Compare with regard to availability, chemical structure, susceptibility to treatment, and eventual utilization in the body, a fat-soluble and a water-soluble vitamin. (London 'A' Level 1973.)

5 The digestive system

In the earlier chapters we learned the body's need for nutrients and which foods supply them. The principal ones, however, such as proteins and carbohydrates, are large complex molecules which cannot pass directly into the blood stream. They must be broken down (hydrolysed) to their simple basic units that are small enough to be absorbed.

Digestion is the process of breaking down complex foods into their simple forms. Proteins are hydrolysed through a series of intermediate compounds into amino acids. Starch is hydrolysed first to dextrin, then maltose and finally glucose. The disaccharides are hydrolysed to monosaccharides (see Table 17).

Part of the dietary fat is hydrolysed to diglyceride and monoglyceride, emulsified with the rest of the fat and absorbed as minute droplets.

TABLE 17
Digestion

Organ	*Function*	*Secretion*	*Action*
mouth	mincing of food	saliva; amylase	starch → maltose
stomach	storage and liquefaction	gastric juice; pepsin, rennin, hydrochloric acid	protein → peptones
liver	fat emulsification	bile; bile salts	assist lipase
pancreas	further digestion	pancreatic juice; lipase, trypsin, chymotrypsin amylase	fat → mono- and diglycerides and fatty acid peptones → polypeptides starch → maltose
Small intestine	final digestion and absorption	succus entericus; peptidase, maltase, lactase, sucrase	polypeptides → amino acids maltose → glucose lactose → glucose + galactose sucrose → glucose + fructose

The vitamins and minerals do not require digesting, they are absorbed unchanged.

Carbohydrates 1. Starch → dextrin → maltose → glucose.
2. Sucrose → glucose + fructose.
3. Lactose → glucose + galactose.
4. Maltose → glucose.

Proteins → peptones → polypeptides → dipeptides → amino acids.
Fats (partially) → diglycerides and monoglycerides.

STRUCTURE OF THE DIGESTIVE SYSTEM (Fig. 19)

The digestive system starts at the mouth where the food is minced by chewing and mixed with the saliva. Starch digestion begins here. When the food is swallowed it passes down the gullet into the stomach, forced along by a series of muscular contractions.

The second stage is the stomach where protein digestion begins. Here the food is mixed with the acid gastric juice by powerful muscular movements until it becomes liquid. Only then can it leave the stomach.

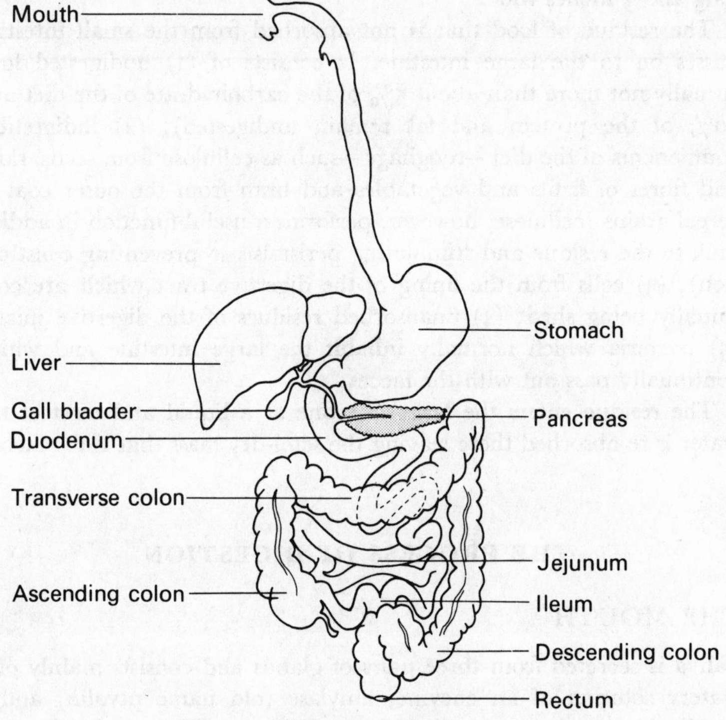

Fig. 19 Diagram of digestive tract

The stomach is separated from the next part of the intestine, the duodenum, by a band of muscle called the pyloric sphincter. This acts as a valve and opens to allow liquid to leave the stomach in a series of gushes.

After the stomach the digestive tract consists of the small intestine (divided into three parts, duodenum, jejunum and ileum) and the large intestine. Three digestive juices are secreted into the upper part of the small intestine, namely pancreatic juice from the pancreas, bile from the liver and a juice from the walls of the small intestine called succus entericus. Digestion of all three major food principles is completed under the influence of these juices and the results of their digestion are absorbed into the blood stream lower down the small intestine.

The small intestine is about 20 feet long and half an inch in diameter. It is made of a double layer of muscle fibres with a lining of mucosa. As the muscular layers contract they push the food along the intestine—rather like pushing toothpaste out of a tube—the process is known as peristalsis. The large intestine is shorter and wider, being about 5 feet long and 3 inches wide.

The residue of food that is not absorbed from the small intestine passes on to the large intestine. It consists of (1) undigested food (usually not more than about 5% of the carbohydrate of the diet and 10% of the protein and fat remain undigested); (2) indigestible components of the diet—roughage—such as cellulose from seeds, skins and fibres of fruits and vegetables and bran from the outer coat of cereal grains (cellulose, however, performs a useful function in adding bulk to the residue and stimulating peristalsis so preventing constipation); (3) cells from the lining of the digestive tract which are continually being shed; (4) unabsorbed residues of the digestive juices; (5) bacteria which normally inhabit the large intestine and which continually pass out with the faeces.

The residue enters the large intestine as a liquid and most of the water is re-absorbed there leaving the semi-dry mass that is the faeces.

THE PROCESS OF DIGESTION

THE MOUTH

Saliva is secreted from three pairs of glands and consists mainly of a watery solution of an enzyme, amylase (old name ptyalin) and a mucilaginous substance called mucin. The amylase has the function of hydrolysing the starch to maltose, and the mucin lubricates the food so that it is easier to swallow.

Saliva is secreted in small amounts continuously and a copious flow is stimulated by the sight, smell, taste, and even the very thought of food. The expression 'makes our mouths water' when we come across particularly attractive food is literally true. Altogether about 1 to 1½ litres of saliva are secreted daily. Savoury, salt and tart foods that we have for *hors-d'oeuvre* and crisp foods like radishes and cucumber, and crusty rolls and thin toast, all stimulate the flow of saliva and help to start digestion.

Although we can detect hundreds of different flavours in foods only four of these can be tasted on the tongue. These are sweetness at the tip of the tongue, sourness at the edges, bitterness at the back and saltiness all over. All other tastes, from fishy to fruity, are detected by the nose. This is why a bad cold makes food seem tasteless. Hot foods have a stronger flavour because the steam carries the aroma to the olfactory organs; iced food needs extra flavouring.

The amylase breaks down the starch first to dextrin and then to maltose but very often food is swallowed so quickly that it is in the mouth for only a short time. The amylase, however, can continue to act in the stomach after the mouthful has been swallowed until the gastric juice eventually penetrates the food and stops amylase action.

THE STOMACH

The act of swallowing is voluntary but once the food enters the gullet involuntary muscular movements, peristalsis, push it down and into the stomach.

Gastric juice is produced by the walls of the stomach and consists of a solution of two enzymes, pepsin and rennin, together with dilute hydrochloric acid. The secretion is stimulated by the presence of food in the stomach and also by the smell and sight of food and its presence in the mouth.

Pepsin, which can only work in the presence of acid, starts the digestion of proteins by breaking them down to soluble peptones. The hydrochloric acid also has a sterilizing action on food which serves to protect us, to some extent, against food poisoning; but it is not very effective if the food is heavily contaminated with bacteria. Nor does it have any effect on the toxins that are produced by food poisoning organisms.

Infected drinking water is a serious hazard to health because water passes very quickly through the stomach before the acid has had time to sterilize it. Even water that is drunk during a meal when the stomach is full of food can leave the stomach within a few minutes by trickling round the stomach wall.

Food cannot leave the stomach until it has become liquid, and the time taken to empty the stomach depends both on the size of the meal and its composition. Soups may leave within half an hour of eating, starchy foods and milk may leave in $1\frac{1}{2}$ hours; on the other hand a large meal may take up to 6 hours before it has all left the stomach; and fat slows down the emptying rate. In general, the stomach acts as a reservoir and enables us to eat at intervals although our tissues are receiving nourishment almost continuously.

A meal that leaves the stomach very quickly is not very satisfying. We may feel hungry by mid-morning after a light cereal breakfast but not after a fatty meal of bacon and eggs.

The function of the second gastric enzyme, rennin, is to clot milk so that pepsin has a chance to attack it. Rennin is obviously very important in children and young animals whose diet consists largely of milk. Rennin from calves' stomachs is sold as rennet for making junkets and curds.

THE SMALL INTESTINE

When the liquefied food leaves the stomach it meets three digestive juices in the small intestine: the pancreatic juice, the intestinal juice, and the bile.

PANCREATIC JUICE

This enters the duodenum through a duct from the pancreas stimulated by two hormones, secretin and pancreozymin, which are liberated from the mucosa lining the small intestine. Pancreatic juice contains three groups of digestive enzymes: (a) protein-splitting enzymes, trypsin, chymotrypsin, exopeptidase, and carboxypeptidase; (b) lipase; and (c) amylase, together with sodium bicarbonate which neutralizes the stomach acid and provides a slightly alkaline medium for the action of the intestinal enzymes.

Trypsin and chymotrypsin are secreted in an inactive form as trypsinogen and chymotrypsinogen and have to be converted into active forms before they can function. They split proteins and peptones into polypeptides. These are then broken down into smaller fragments, tripeptides and dipeptides, by exopeptidase and carboxypeptidase.

(The pancreas also secretes the hormone insulin directly into the blood-stream but this has no connection with digestion.)

INTESTINAL JUICE (SUCCUS ENTERICUS)

This is secreted from glands in the walls of the small intestine and consists mainly of a special enzyme enteropeptidase (enterokinase), which is not directly connected with digestion. Its function is to convert the inactive trypsinogen of the pancreas into trypsin. The chymotrypsinogen is activated by trypsin itself.

The final phase of digestion takes place inside the epithelial cells of the lining of the small intestine—called the brush border. The cells contain dipeptidases which split the small peptides into individual amino acids and also disaccharidases to hydrolyse maltose, sucrose, and lactose. A small amount of these enzymes appears in the small intestine in the intestinal juice but they have no real function there.

BILE

Bile is formed in the liver and stored in the gall-bladder until it is needed. It contains bile salts (made from cholesterol), lecithin, and cholesterol and is required to help to digest fats.

Fat digestion is somewhat complicated. Part of the triglycerides is successively hydrolysed by the lipase of the pancreatic juice to a mixture of diglycerides, monoglycerides, glycerol, and fatty acids. All these, together with some unchanged fat, are emulsified with the aid of bile salt to form very small droplets called micelles. So both pancreatic juice and bile are required for fat digestion.

These micelles are absorbed into the cells lining the intestine and they carry with them the fat-soluble vitamins.

Inside the cells they undergo further change. Some of the monoglycerides are hydrolysed completely to glycerol and fatty acids and then the glycerol is linked to long-chain fatty acids (i.e. it is re-esterified) to form new types of triglycerides. They now no longer resemble the dietary fats but have been changed into fats characteristic of human beings.

ABSORPTION

The walls of the intestine provide an enormous surface area since not only is the mucous membrane puckered into numerous folds and ridges but it is also covered with vast numbers of small hair-like projections called villi (Fig. 20). These are covered with microvilli called the brush border and it is inside these cells that the final stages of hydrolysis take place.

The villi are traversed by two systems of capillaries, (a) blood capillaries through which monosaccharides and amino acids are absorbed and (b) lacteals through which fat is absorbed.

Blood from the capillaries is collected into the portal vein which takes the nutrients to the liver.

The triglycerides that have been re-formed in the cells enter the lacteals as small droplets called chylomicrons and enter the bloodstream later. Two to four hours after a meal rich in fat the blood looks milky because of these droplets, which can easily be seen under the microscope. They soon go to the fat storage depots, or to the muscles to be used as fuel, and the blood clears.

Triglycerides with carbon chains of 12 atoms or less can be absorbed directly without undergoing the emulsification process. They are more easily hydrolysed by pancreatic lipase into glycerol and medium-chain fatty acids, and these can be absorbed directly into the blood capillaries of the villi. Consequently they can be absorbed by people who suffer from defects of fat digestion or absorption and it is possible to obtain these 'medium-chain triglycerides' for such patients by separating them out from ordinary fats which contain both medium-chain and long-chain triglycerides.

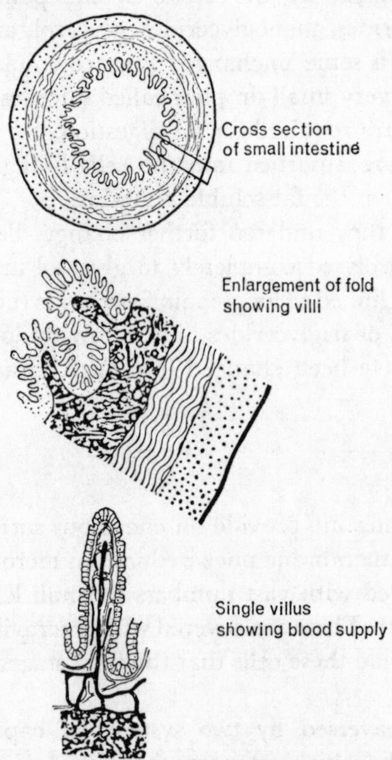

Fig. 20 Absorptive area of the small intestine

Normally 90–95% of our food is digested and absorbed. When there is a blocked bile-duct, as in obstructive jaundice, or lack of one of the enzymes, or damage to the intestinal mucosa then digestion and/or absorption are impaired and part of the food passes straight through the intestine (malabsorption). For example, there are some people who lack the enzyme lactase, at least after childhood, and while they can tolerate small amounts of milk, larger amounts cause diarrhoea because the lactose passes down to the large intestine. In coeliac disease there is damage to the villi of the small intestine caused by the wheat protein, gliadin, and much of the food fails to be absorbed.

LARGE INTESTINE

The 5–10% of the diet that is not digested includes plant materials like cellulose, hemicellulose, lignin, gums, and pectin. Together with large volumes of fluid from the digestive juices this travels along the intestine —much of the water is re-absorbed in the large intestine leaving the residue in a semi-solid state.

The large intestine contains large numbers of bacteria which can flourish on undigested food residues. For example, one of the reasons for the flatulence that follows eating some kinds of beans is that a small amount (2 or 3%) of the carbohydrate of the beans consists of sugars (raffinose and stachyose) that are not digested but travel down to the large intestine where the bacteria metabolize them and produce gases—methane, carbon dioxide, and hydrogen.

Dietary fibre and its connection with various disorders of the bowel is discussed in Chapter 6, p. 103.

PRACTICAL CONSIDERATIONS

Unless one is suffering from some disorder that affects the digestive enzymes or the bile or the intestinal mucosa, most of the food that we eat, 90–95%, is digested and absorbed, but the time taken can be prolonged. This causes great discomfort which is popularly called 'indigestion'.

Emotion, for example, affects the rate of emptying of the stomach. In a state of fear or emotional disturbance food can remain in the stomach for 12 hours or more, whereas excitement can speed up emptying.

The sight and smell of food help to stimulate the flow of gastric juice, which means that attractively prepared food served in pleasant surroundings aids the process, while unappetizing food really means what the term says.

Certain savoury foods like meat extracts provoke a gastric secretion when they reach the stomach. These are the kinds of food often eaten as hors d'oeuvres or which may be present in soups. There is also some effect on other digestive juices and on the muscular tone of the stomach and small intestine.

We could eat much of our food raw, and some people do, but generally the digestive enzymes cannot easily penetrate raw foods like cereals, roots, and pulses, and tough meat is difficult to chew. Some part of such raw foods, especially if swallowed without being properly chewed, may escape digestion and pass down to the large intestine where it can cause flatulence.

So it is obvious that the cook can greatly affect our appetite, comfort, and sense of well-being after a meal, even when there is no effect on the amount of nutrients actually absorbed. The appearance on the plate, including colour, arrangement, and garnishes, textures such as crispness and tenderness, as well as smell and presentation and surroundings can help or delay the digestive processes.

AFTER ABSORPTION

All the carbohydrates of our diet ultimately end up as glucose. During digestion glucose, fructose and galactose are formed, and there may be small quantities of the pentose sugars present in the food. All these sugars are converted into glucose in the liver, and it is glucose that is the 'blood sugar', the carbohydrate that is used by the tissues of the body.

If glucose is consumed it does not, of course, need digesting. This is one of the reasons why glucose is added to many foods—it is regarded as a rapidly available form of energy. In fact sucrose, and even starch, are hydrolysed so quickly that they reach the blood stream (or at least the glucose formed from them reaches the blood stream) almost as quickly as free glucose does.

When the nutrients are absorbed from the intestine they pass into the portal vein which travels first to the liver. Here some of the glucose is removed and converted into glycogen which is stored in the liver until it is needed. The remainder passes on to the muscles where it is used for energy or stored as muscle glycogen. In a well-fed person there is about 250 g of glycogen in the liver and another 100 g in the muscles (a total of about 1 300 kcal). The general pattern is that after a meal the blood sugar level (which rises to 140 to 150 mg per 100 ml) is kept down by removing the glucose and storing it as glycogen; between meals (when the blood sugar can fall to 80 to 90 mg) the level is kept up by converting glycogen to glucose.

The fat, after absorption, also travels to the liver where it may be oxidized for energy, pass to the muscles for oxidation, or travel to the adipose tissues where it is stored as body fat.

The amino acids travel to the various tissues where they are rebuilt into whatever type of protein is needed, such as enzymes, hormones, muscle and tissue proteins. There is no storage of amino acids in the body and any surplus to immediate needs is oxidized.

EXAMINATION QUESTIONS

1. Explain the processes involved in the digestion of raw cheese. Of what use are the constituents of cheese to the body? (City & Guilds Adv. Dom. Cookery.)
2. What is the reason for serving *hors-d'œuvre* at the beginning of a meal and what are the main points to consider when planning them?
3. Trace the pathway of digestion and absorption of a light meal of mushroom omelette, watercress, and fresh orange juice. (London 'A' Level 1975.)

4. Explain fully why it is necessary that students of Cookery should know about the processes of digestion. (City & Guilds Adv. Dom. Cookery.)
5. Discuss the importance of protein in the diet. What are the main sources of this nutrient? Describe the processes involved in its digestion. (City & Guilds Adv. Dom. Cookery.)
6. What factors (a) stimulate a flow of salivary and gastric juices, (b) promote good digestion of a meal? Of what use are these facts to a caterer planning (a) midday meals in a works canteen, and (b) lunches in a city restaurant at a moderate cost? Illustrate your answer with two menus. (R.S.H.)
7. What digestive processes are performed by (a) saliva, and (b) pancreatic juice? Describe an experiment to illustrate the (former) function of saliva. (R.S.H.)
8. Give an account of the digestion and uses of fat in the body. Compare the nutritive values of butter, cooking fat and vegetable oil. (R.S.H.)
9. What nutrients are present in milk? Describe how one of these nutrients is digested, absorbed, and utilized by the body. (Hotel & Catering, Intermediate.)

10. Why is it important to include dietary fibre in our diet? How can this be provided in the diet of elderly people?
11. What are the main sources of dietary fibre in (a) English diets, (b) Asian diets? What are the results of an inadequate intake of dietary fibre? How would you encourage housewives to use bran in meals prepared for their families? Illustrate your answer with two recipes.
12. Discuss the importance of fish and chips in the British diet. Briefly describe how the main nutrients are digested. (RSH Certificate 1975.)

6 Nutritional problems in the 'richer' countries

There are some known deficiencies in the diet even in 'rich' communities. Some degree of anaemia due to a shortage of iron is common, particularly among women, in *all* areas, and so iron is added to the bread in many countries, although there is considerable argument as to how serious the problem is.

There also appears to be a marginal deficiency of certain vitamins, such as folic acid and vitamin B_6, which shows up more particularly in pregnant women.

There is also a problem of rickets in Great Britain, both among Asian immigrants, whose dietary habits differ from the rest of the population, and to some extent among native-born adolescents.

One of the causes of dietary deficiency amidst an abundance of food is poor selection of foods.

SECTION A DISEASES OF AFFLUENCE

In recent years a number of diseases have become widespread in the industrialized countries of the world, which have been called 'diseases of affluence'. They include coronary heart disease, diseases of the bowel, various forms of cancer, and dental decay. There are very many causes of all of these and it is not certain that diet is even one of the causes, but it does appear to be involved somehow. Since the diet is often one of the very few factors that can be altered, much research, still inconclusive, has been devoted to the role of the diet in these disorders.

CORONARY HEART DISEASE (ischaemic heart disease)

One of the most intensive areas of nutrition research in recent years has been into the possible relation between diet and heart disease. Heart disease has increased at such a rate during the last generation that it has become one of the major causes of death in men over 40 years of age; many factors are known to be implicated—genetic factors, smoking, blood-pressure, lack of exercise, stress, diabetes, and possibly diet.

The subject is extremely complicated and can be only very briefly summarized here.

The disease is caused by an accumulation of a material on the lining of the coronary arteries supplying blood to the heart muscle itself. This consists of fatty substances, complex carbohydrates, and fibrous tissue called atherosclerosis. The deposit narrows the space in the arteries and reduces the flow of blood to the heart muscle. The disease is far commoner in the industrialized communities than in developing countries.

The risk of heart disease is greater when blood levels of certain kinds of fats, including cholesterol, are higher than average, and the average risk increases with higher levels. It is important to note that these are observations of *averages* of population groups and that there are plenty of people with high cholesterol levels who do not get heart attacks and vice versa. Moreover, different types of blood fats might be increased and they respond in different ways to dietary changes. Consequently much research has been devoted to finding methods of reducing levels of fats in the blood and of ascertaining whether such reductions are beneficial. The evidence is still not clear.

All three fatty substances in the diet, the saturated fats, the polyunsaturated ones, and cholesterol, are thought to be implicated in some way in the problem of heart disease. Since most (not all) of the saturated fats are animal fats, and most (not all) of the vegetable oils are rich in polyunsaturated fats, there is a distinction often drawn between animal fats and vegetable oils, but the real distinction is in the kind of fatty acids present.

First, there is cholesterol in the diet. The body manufactures its own cholesterol and moderate intakes in the diet simply reduce the amount made and so do not affect blood levels. The average Western diet contains about 500–1000 mg of cholesterol a day coming from eggs (about 300 mg each), brain, and to a lesser extent liver. A piece of *general* nutrition advice (which might not matter at all to some people with their particular pattern of blood fats,) is to keep their dietary cholesterol down to about 300 mg per day.

Second, there are simple fats. The increase in coronary heart disease appears to run parallel with the amount of fat that we eat. This has increased from about 35% of the diet to over 40%—measured as a percentage of the energy intake rather than a per cent of weight of food—over the past forty years. At the same time people in developing countries where fat is a much lower proportion of the diet, as little in some areas as 10–20%, rarely get heart disease. Consequently it is believed that a high fat diet is one of the factors.

Thirdly, there is the type of fat. It is possible to reduce the level of cholesterol in the blood by consuming the polyunsaturated fats. These are mostly present in vegetable oils, especially corn (maize), sunflower and safflower oils, and, of course, margarines made from these. It is also found by experiment on human subjects that cholesterol levels are increased by some of the saturated fatty acids found in the animal fats. So it is suggested, although not fully accepted, that the type of fat in the diet should be changed.

Fourthly, other dietary factors have been implicated, including the high consumption of sugar relative to starch in countries where heart disease is common, the low intake of dietary fibre, and the unexplained observation that there is a lower death rate from heart disease in areas where the drinking water is hard.

So while it has not been proved that diet is a factor in the causation of heart disease there is considerable evidence that the diet plays some part.

Dietary experiments have been carried out on large numbers of people in which saturated fats have been partly replaced with polyunsaturated ones and blood cholesterol levels have been lowered as a result, but the changes have had no clear-cut effect on the incidence of heart disease. This is why intensive research continues. Meanwhile if it is true that diet is a contributory factor, then it is possible to take steps to improve our diet.

Dietary advice falls under two headings: (1) to those known to be at risk or who have already suffered a heart attack, and (2) general nutritional advice to the public. The former would receive specific advice related to their particular type of blood fat disorder. The latter has led to official nutrition policies in some countries, although not everyone agrees with all the recommendations.

1. Avoid obesity.
2. Reduce the fat intake from current levels of 40% to 30–35%.
3. Reduce the intake of saturated fats (which largely come from meat) and increase polyunsaturated fats.
4. Reduce dietary cholesterol to 300 mg per day (3 eggs per week).
5. Reduce sugar intake.
6. Increase the consumption of fruits, vegetables, and cereal products.

The recommendations were expanded and published by U.S.A. Select Committee (1977) as 'Dietary Goals' (see Appendix).

BOWEL DISEASE AND DIETARY FIBRE

One of the methods of finding the relation between diet and disease is to study the disease patterns and the eating habits in different communities. This is termed epidemiology and has led to two findings in this area. First, a number of diseases have all increased considerably in recent years in Western communities and hardly ever occur in developing countries. This suggests a common cause for them. Second, they occur in communities where the intake of dietary fibre (undigested roughage) is low.

It is consequently suggested, although not universally accepted, that a low intake of dietary fibre is the cause of these diseases. Certainly many people have relieved constipation and consequent disorders by eating whole-grain cereal such as wholemeal bread or bran, and doctors have alleviated diverticular disease in the same way, but whether low fibre intakes are the cause of the disease is still under investigation.

The difference in diets is that many communities prefer to eat white flour and sugar (brown sugar and white are almost identical) while more primitive peoples and ourselves in earlier generations obtained their starch as whole-grain cereals and their sugar as fruits. Hence the criticism of 'refined carbohydrates'.

The term 'dietary fibre' refers to the undigested part of plant foods including cellulose, pectin, hemicellulose, lignin, and a variety of complex carbohydrates. It is cereal fibre, i.e. bran, that appears to be most beneficial.

The terminology has become confused because analysts for many years have measured 'crude fibre' by finding what was left after treatment with acid and alkali under specified conditions of analysis. This bears very little relation to dietary fibre (which is, in any case, not a fibre) but most of the analyses in older food composition tables carry figures of crude fibre.

The amounts recommended in the diet vary with the individual but average *about* 5–6 grams of dietary fibre or 20 g of bran or 4–5 slices of wholemeal bread per day.

So the consumption of whole grain cereals should be added to the dietary goals listed above.

SECTION B VULNERABLE GROUPS

Good health can be easily maintained in the mature adult so long as he satisfies his appetite and his foods are varied. There are, however, certain groups of the population, sometimes called the vulnerable groups, whose diet needs rather more attention. For example, since babies and young children are growing rapidly they require higher proportions of certain nutrients than adults do. Pregnancy is an obvious instance where the mother has to consume enough protein, minerals and vitamins to supply the baby as well as herself, while the nursing mother has an even greater need for all nutrients. Convalescents, as described in Chapter 4, have an increased need for protein and also vitamins.

THE ELDERLY

Elderly persons are an especially vulnerable group. As people grow older they spend less energy because their basal metabolism decreases and they spend less effort in various activities than they did when younger. They will therefore need less to eat.

The F.A.O. Committee recommends the following reductions in energy needs with advancing years. For the decade 25 to 35 and 35 to 45, a decline of 3% each decade compared with the needs at 25 years of age; for the decades 45 to 55 and 55 to 65 a decline of 7·5% per decade; and from the age of 65 to 75 a further decline of 10%. The total energy expenditure of persons of 75 can, therefore, be only two-thirds of what it was when they were 25 (see Fig. 12); therefore, their needs are only two-thirds.

This means that they will be consuming two-thirds of the protein, vitamins and minerals, if they are eating the same type of food.

It is important to realize that while their energy needs are certainly reduced, their need for vitamins, minerals and protein is *not* less than it was. Moreover, many elderly persons, because of infirmity, apathy and reduced sense of taste tend to eat less food, and very often that food is less nutritious. Tea and biscuits are often their standby. There is, therefore, some risk that they may be short of proteins, minerals and vitamins. In particular, many elderly people are anaemic and so should have iron-rich foods, and may suffer from osteomalacia (see page 79) and would benefit from calcium and vitamin D.

Consequently, elderly people should eat more of the good quality foods, i.e. foods such as milk, cheese, eggs, liver, meat and fish, which contain a higher proportion of these nutrients, as well as adequate amounts of fruit and vegetables. Since the appetite may be poor there is no room in the diet for many of the sweets and puddings that are largely a source of energy and little else.

Planned meals When planning meals for old people in hospitals, at home or for the 'Meals on Wheels' service and other welfare schemes, the caterers should aim at providing 50 g of good quality protein per day. This could be done, for example, with 1 pint (570 ml) of milk, 4 ounces (114 g) sausages and a 6 ounce (170 g) lamb chop (using the list of protein foods on page 68), or, more cheaply, by using less meat and making use of concentrated protein foods such as dried skim milk powder, soya flour, cheese and dried brewers' yeast. 1 ounce of dried milk could be mixed into a milk pudding, drink, blancmange or soup; 1 ounce (28 g) of cheese could be added to pastry, mashed potatoes, sauces, or served grated for sprinkling on to soups and vegetables.

Old people like bread and sandwiches because they are easy to eat. A variety of nutritious spreads can be made with the aid of a liquidizer or mortar such as paté (liver, eggs and fat), kipper spread (kippers, cheese and margarine), savoury spread (dried milk, soya, brewers' yeast or yeast extract and fat), as well as the more usual meat and fish pastes.

Old people *need* foods of concentrated food value, and they *like* foods that are easy to prepare and eat, and especially if they are tasty. Bland, delicate flavours are popular with children: they may, for example, like cream cheese but many older people prefer gorgonzola.

A suitable meal for an old person would be fish pie, made by poaching filleted fish in white sauce with a savoury crust of cheese and mashed potato, served with tomato juice with a dash of Worcester sauce, followed by fruit in jelly or stewed fruit (with added orange juice for extra vitamin C) and chocolate sauce.

An unsuitable meal would be a small portion of fried fish with a large helping of chips, followed by a syrup pudding or jam tart with custard. This would be rich in energy and poor in protein, minerals and vitamins. But it is quite likely that the unsuitable meal would be preferred by both old people and caterers alike—cheaper to the latter and more akin to the old food habits of the former.

The second meal could be improved with more fish and less chips and by substituting a sweet such as egg custard or fruit to cut out the sugar and starch.

Where old people cater for themselves they should be advised to buy cheaper protein foods such as fish roes, coley, cheese and dried skimmed milk, and simple recipes should be suggested. Drinks could be made from dried skimmed milk mixed with cocoa or with savoury meat and yeast extracts; soups can be made up with the skimmed milk powder mixed with water. Dried skimmed milk keeps well and is easy to use.

The extra effort devoted to meals for elderly people is very well worth while as a great deal of the ill-health which is so common among elderly people can be alleviated by an improvement in their diet.

BABIES

The feeding of babies in the first few months is relatively simple since milk supplies nearly all their needs. The only additions necessary are vitamin C, supplied as fruit juice, and vitamins A and D, supplied by cod liver oil or similar preparations. If the baby is bottle-fed attention must be paid to mixing the correct proportions of milk and water, and to hygiene. Babies are not as resistant to infection as adults and greater care must be taken to see that their food and utensils are clean and, where necessary, sterilized.

When the baby reaches 3 to 4 months or, a better criterion, about 13 pounds (6 kg) in weight, milk no longer supplies all his needs and other foods must be added to the diet. One vital need is for iron. At birth the baby has a store of iron sufficient, if the mother was well fed, to last for 3 to 4 months. Both breast milk and cow's milk are deficient in iron (although it is added to most proprietary dried milks) and so it is necessary to include some food rich in iron in the diet at this age, e.g. minced liver, puréed apricots, broths and gravy. The inclusion of foods other than milk in the diet allows the baby to get used to the taste and texture of adult foods.

Although only a generation ago it was thought that solid foods should not be given to babies until they were at least a year old, it is now common practice to give a variety of foods at 3 months and even earlier. Babies can eat almost any of the foods eaten by the rest of the family so long as they are minced or sieved. The only foods that are not acceptable are the strongly spiced ones such as pickles, sauces, and curries.

Meat, fish, eggs, cereals, vegetables and fruits, in the strained or minced form, are gradually added to the diet as the baby reaches 6 to 12 months of age. These foods can be prepared at home using a liquidizer or grater; but great care must be taken to ensure that the utensils are kept clean. They can be bought ready-made in tins or bottles or in the dried, powdered forms.

In his first year the baby will probably be taking more than a pint of milk a day plus the various additions; and the amount of milk is reduced somewhat in the second year as more of the ordinary foods are eaten. However, a large amount of milk in the diet throughout childhood is an excellent method of ensuring a good diet.

Children often have fads and fancies which result in some degree of restriction of the diet. The guiding principle is that appetite will almost always fulfil the energy needs but attention must be paid to the other nutrients. If one protein or vitamin-rich food is refused then it should be replaced by another one of similar food value: e.g., if cooked greens are unpopular, fruit juices or raw vegetables can replace them; if meat or fish is disliked, cheese or eggs are a useful substitute.

PREGNANCY AND LACTATION

The expectant mother obviously has special dietary requirements. She has to consume enough protein and minerals to construct the body of the baby, as well as the energy and vitamins that are also needed. Although there is relatively little growth of the foetus during the first 6 months of pregnancy it is important that the mother should store, insofar as the body can store them, enough nutrients to supply the high demand that will be made on her tissues during the last three months.

A high intake of calcium and iron is particularly important for the formation of bones and blood. While the demand for calcium can be met if an extra pint or so of milk is taken daily (and this will also take care of the needs for protein and vitamins A and B_2) the iron intake is not so easily ensured. Meat, liver and fortified cereals can make some contribution but it is usual to ensure that there is enough iron in the diet by taking tablets of iron preparations. Tablets or capsules of vitamins are another useful insurance to meet the increased demands.

During lactation the mother's needs are even greater since she now provides nourishment for a much larger baby. The tables show the recommended intake levels at this stage: protein and vitamin-rich foods, and particularly foods rich in calcium, are of great importance in the diet. It is worth noting that the nutrient content of the mother's *milk* is not much affected by dietary shortages, but it is the mother herself who suffers. The quality of the milk is generally maintained and if the mother's diet is not adequate the shortage is made good from her own body stores.

ADOLESCENTS

In adolescence there is a sudden spurt of growth resulting in greater height and stronger bones and muscles. This rapid growth can take place at any time between the ages of 12 and 19 years; it creates exceptionally high needs for energy, greater than those of the adult, see Table 6. A good appetite usually ensures that enough energy foods are taken and so it is likely that enough protein is also taken; but there is a danger of filling up on 'empty calorie' foods such as sweets, iced lollies and sweet drinks. The caterer should, therefore, make food of good nutritional quality available such as milk drinks, fruit juices, fruits, cold meats, cheeses in variety, bread, buns, and yeast cakes, fish and savoury dishes, together with novelties such as pizzas. The less valuable foods, such as meringues and sweet cakes should be dearer than the more nourishing foods so that they are eaten as an extra and not a replacement of the better foods.

INDUSTRIAL FEEDING

People differ considerably in their eating habits: some eat little or no breakfast; some have only one, others two, large, meals a day. In general, so long as enough is eaten over the whole day, there appears to be no harm in these different habits of eating. Where a day's work has to be done, however, more attention has to be paid to the spacing of meals. For example, it has been shown that people who have no breakfast are below par by the middle of the morning, as shown by their impaired ability to perform certain mental and physical tasks. A small breakfast of about 300 kcal makes a difference. (A larger breakfast does not produce any further improvement.)

What is needed, therefore, is to ensure that something is provided in the early morning even if it is only a light meal. This could be a sandwich and hot drink prepared overnight and kept hot in a vacuum flask, or a mixture of cereal and raisins with milk, and a drink of orange juice, or an egg and toast. Where people travel a considerable distance to work and do not feel inclined to eat an early breakfast, it would be practical to provide a light meal on arrival or at an early 'break'. Milk or a hot drink, together with a sandwich or a cheese roll would ensure a better morning's work in schools, offices and factories.

Again, at lunch time, it is necessary to ensure that a nourishing meal is eaten if an efficient afternoon's work is expected. The traditional meal of meat and two 'veg' with pudding and custard to follow is not to everyone's liking in the middle of the working day. It is apt to cause sleepiness in sedentary people and indigestion in the active ones. A smaller or more concentrated snack meal may be preferred, such as toasted sandwiches, salad plates, cheese boards, open sandwiches, minestrone-type soups containing pasta, vegetables and meat served with a roll and cheese. Yoghurt, fruit and ice-cream desserts can be more nourishing than suet puddings. Calculations given on page 27 show that snack meals, if properly chosen, can have as high a food value as the conventional meal. If plain filling foods, such as rolls, toast, jacket potatoes, bread or potato salad are served *ad lib.*, then people will be able to take enough food to satisfy their individual energy needs.

These are matters of some importance to maintain health and efficient working. If good attractive food is not provided at the place of work then the alternative is that people will go elsewhere and may subsist on unsatisfactory food of poor quality.

SLIMMING

When excess vitamins B and C are eaten, the surplus is excreted; but when excess protein, fat or carbohydrate is eaten the surplus stays in the body as fat. It is usually stored under the skin and around the less active muscles. Fat people have eaten more food than they need and the remedy is to eat less and use up the stored fat.

There is some danger that when fat people reduce their food intake they may consume insufficient protein, vitamins and minerals, hence slimmers, especially teenage ones, can become a nutritionally vulnerable group. The aim is to reduce the energy intake without allowing the intake of proteins, minerals and vitamins to fall below the recommended levels.

Weight control and the lifelong possession of a slim figure depend on balancing energy intake with energy spent on activity. In order to reduce weight, cut out the sugary-starchy foods which supply chiefly calories, such as sweets, icings, meringues, puddings, crisps, biscuits, cakes, pastry and, particularly, sugar itself. Sugar provides an average of about 25% of the calories of the British diet and if this were halved, not for a short period while slimming but permanently, then overweight with all the ill-health that this causes (diseases such as diabetes and heart disorders) and certainly dental decay would be much reduced.

Many people find it rather difficult to eat less—they like their food too much—but exactly the same effect is achieved, without feeling hungry, if the carbohydrates are severely restricted, while still continuing to eat as much protein and fat as you like. This does not mean simply cutting out sugar, or eating starch-reduced bread. It means a severe restriction on all carbohydrate-containing foods. The details are available in Professor John Yudkin's book *This Slimming Business*, Harmondsworth, Penguin (1965).

Weight loss One important fact must be borne in mind when embarking on a slimming diet—body fat is never lost quickly. Whether the diet is based on drugs that depress the appetite, or formula diets, or emphasis on low-calorie foods, changes in weight due to loss of body fat will always be slow. One pound of body fat is equivalent to 3 500 kcal. If you cut your energy intake from 2 500 to 2 000 kcal a day, that is by a fifth, you would lose only 1 pound of fat a week. Even if you ate *nothing at all* you could only lose about 6 pounds in a week! Those methods that result in a quicker loss of weight achieve results by causing a temporary loss of water but not body fat, e.g. Turkish baths.

Special slimming foods and appetite depressants may be of value in encouraging the slimmer but the sad fact is that he will go back to his original size as soon as his normal, bad, eating habits are resumed. Weight control calls for improved eating habits and they must be maintained for life.

Weight control To make sure that good nutrition does not result in over-nutrition the following rules may be helpful.

1. Eat less of the carbohydrate foods, especially sugar.
2. Eat more fruit and vegetables—which are poor in energy.
3. Eat plenty of protein foods such as cheese, milk, eggs, fish, and meat.
4. Eat smaller meals more frequently rather than one or two excessively large meals in the day—and that means don't go without breakfast.
5. Spend more of your energy by taking up some activity, such as walking or games.

Today, when so many people are wisely trying to regain their proper body weight, caterers have a responsibility to provide the right foods. (See Table 37, p. 191.)

EXAMINATION QUESTIONS

1. You are put in charge of the production and distribution of cheap or free meals to elderly people in your area. Describe in general terms your approach to the problem. (R.S.H.)
2. You are responsible for the catering in a convalescent home for post-operative patients. With what nutritional principles in mind would you plan the meals? Illustrate your answer with menus for 2 days. (R.S.H.)
3. Plan midday meals for one week for a school canteen. What factors would you consider particularly important in making the menus? Give the quantities of foods used for one of the meals, planned for 100 primary children. (R.S.H.)
4. Plan a day's menu for an old lady with a small appetite. How would you ensure that she had sufficient nutrients to supply her needs? (R.S.H.)
5. Plan main tea meals for 5 days suitable for a 4-year-old child, giving the reasons for your choice of foods. (R.S.H.)
6. Explain how you would provide adequate nourishment for an elderly couple living on a small income. Plan their meals.
7. How does a knowledge of cookery and food values ensure adequate nutrition when planning diets for (a) someone who wishes to slim, (b) a person who has difficulty in digesting fat, (c) an

active teenager? Illustrate your answer with a menu of a day's meals in *one* of these cases. (City & Guilds 244.)

8. It is not always practicable to provide a diet which contains the recommended amounts of all the nutrients every day. Discuss this and state what fluctuations you would tolerate in respect of (a) vitamin C for inmates of an old persons' home, and (b) protein in a residential college for students. (R.S.H.)
9. What are the normal dietary needs of a woman over 20 years of age in a sedentary occupation? How are these altered during pregnancy and lactation? (R.S.H.)
10. You are asked to help at an infants welfare centre by showing a group of young mothers how to feed their children aged 3 to 5 years. List the nutrients which children especially need. Make a display of dishes and meals suitable for this age group. Write display cards showing how much each dish would contribute to their daily nutritional requirements. Calculate the energy and calcium value of one dish. ('A' level practical Home Economics.)
11. You are asked to prepare a buffet lunch for a canteen to launch a slimming project for women staff. There should be special emphasis on low carbohydrate, and high protein foods and also low energy-value foods. The menu should include a selection of hors d'oeuvres, soup, choice of 2 main dishes to be served with green salad, 2 dishes using fruit and 2 using cheese. Present your test with appropriate notices giving the approximate energy value per portion. Grade the foods for carbohydrate value using 1, 2 or 3 stars in descending order of value. ('A' level practical Home Economics.)
12. Teenage girls at a city school for 700, rejected the school meal of meat, two vegetables and pudding, and preferred to buy chips, hot dogs, 'burgers, baked beans and fish fingers, with sweet cakes and lemonade at local cafés. The school canteen then provided these foods and many more girls stayed to the school lunch. Comment on this situation for the nutritional and social points of view and describe how you, as school meals supervisor, would provide a suitable meal in these circumstances.

7 Commodities

We eat foods, not nutrients. In previous chapters, the nutrients have been studied—their functions, sources and the amounts needed to provide the energy and materials of construction of the body. In this chapter, this knowledge will be used to evaluate foods and to help caterers to make a sound choice when planning menus and diets.

Food choice depends on conscious and unconscious influences. We say 'we know what we like', but it would be more accurate to say 'we like what we know'. The foods we choose result from the interaction of such factors as habit, cost, taste, taboo, class, season and storage, availability and ease of preparation.

To make a good choice of foods it is necessary to know their composition and then to evaluate them by calculating the proportion of the daily allowances of the various nutrients supplied by an average helping. Deficiencies in one food can be made good by mixing with other foods, either in the recipe or in the meal, and so a balanced diet is obtained, i.e. one supplying all the nutrients needed.

An occasional unbalanced meal does not matter; but badly balanced types of meals occurring regularly, either as a result of poor food habits, or the caterers' mistaken choice, add up to ill-health and waste of time and money.

Many foods have a similar composition, and can be studied in groups, e.g. cereals, fish, meat, eggs, milk, starchy roots, etc.

To study the nutrient value of a food or group of foods:

(a) note the composition per oz or per 100 g of the edible portion, and the amount of waste;

(b) decide the size of the average portion eaten, or supplied by the caterer;

(c) calculate the proportion of daily allowances of nutrients supplied by this portion;

(d) note the chief value of the food and its deficiencies;

(e) consider what other foods should be consumed at the same meal to make good the deficiencies; and,

(f) consider digestibility and effects of food preparation; take note of significant points such as keeping quality, etc.

Note. The composition of foods can differ to some extent from one sample to another so that the figures given in different food composition tables do not always agree. The figures used in this chapter are taken from Platt, B. S., *Tables of Representative Values of Foods Commonly Used in Tropical Countries.* Medical Research Council Special Report Series No. 302, London, HMSO, and McCance, R. A. and Widdowson, E. M., *The Composition of Foods.* Medical Research Council Special Report Series No. 297, London, HMSO (revised Paul and Southgate, 1978).

MEATS

Meats are chiefly valued for their satiety value, the central place they occupy in a meal, and their contribution to the protein, B vitamins and iron of the diet.

An average portion of beef, pork and lamb is 3 oz (= 3½ oz with waste). An average portion of poultry is 4 oz (= 6 oz with waste).

TABLE 18
Average composition of lean meats
per oz (28 g) edible portion

	Waste (%)	*kJ*	*kcal*	*Protein* (*g*)	*Fat* (*g*)	*Fe* (*mg*)	B_1 (*mg*)	B_2 (*mg*)	*Nic.* (*mg*)
Bacon	12	400	95	3·6	9	0·4	0·16	0·05	0·8
Beef	23	220	53	5	4	0·8	0·03	0·05	1·3
Mutton	26	160	40	4	2·5	0·7	0·05	0·06	1·3
Pork	18	420	100	4	9	0·5	0·2	0·05	1·0
Poultry	33	150	35	5	2	0·4	0·03	0·04	2·5

TABLE 19
Food value of average portion
A per portion (3 oz (85 g)); *B* expressed as fraction of adult allowance

		MJ	*kcal*	*Protein* (*g*)	*Fe* (*mg*)	B_1 (*mg*)	B_2 (*mg*)	*Nic.* (*mg*)
Daily allowance		11	2 500	50	12	1·0	1·5	18
Beef	*A*	0·7	160	15	2·4	0·10	0·15	4
	B		1/15	1/3	1/5	1/10	1/10	1/6
Mutton	*A*	0·5	120	12	2·0	0·15	0·18	4
	B		1/20	1/4	1/5	1/7	1/8	1/6
Pork	*A*	1·3	300	12	1·5	0·6	0·15	3
	B		1/8	1/4	1/8	1/2	1/10	1/6
Poultry	*A*	0·6	144	20	1·6	0·12	0·16	9
4 oz (114 g)	*B*		1/20	2/5	1/8	1/8	1/10	½

FOOD VALUE

Meats are good sources of protein and nicotinic acid, and useful sources of iron and vitamins B_1 and B_2. Meat proteins are good quality, especially useful to supplement cereal proteins which are relatively poor in lysine. In countries with diets composed of 70 to 80% cereal, the addition of small quantities of meat greatly increases the protein quality of the whole diet. Pork is especially rich in thiamin.

Meats are poor sources of calcium, carbohydrate, and vitamins A, C and D.

FOOD PREPARATION

Ultimately both tender and tough meats are digestible *if* they can be equally well chewed. There is no difference in nutritional value between the cheaper, tough cuts and the more expensive tender cuts.

Acidity tenderizes meat proteins: e.g., as when acids develop on hanging, or using marinade. Tenderness is related to length of meat fibres. Meat can be tenderized by cutting across the fibres to shorten them, also by the enzyme papain (sold as meat tenderizer).

Proteins coagulate at 170 °F (77 °C) and harden at higher temperatures. *All* meats become tender if cooked at 275 to 300 °F (130 to 150 °C) for 5 to 8 hours, or until a thermometer stuck in the thickest part registers 170 °F (77 °C). Good quality meats can be cooked at higher temperatures, 400 °F (205 °C) for a shorter time. B vitamin values are reduced by heat (30% thiamin loss) and by solubility in juices (10% loss of riboflavin and nicotinic acid).

BALANCED RECIPES

The nutrients poorly supplied by meat can be added from foods such as milk and cheese (calcium and vitamin A), margarine (vitamins A and D), bacon, kidney, flour (thiamin), vegetables (vitamin C), cereals, and potatoes (carbohydrate); e.g.

(a) Spaghetti bolognaise (tomato, ham, meat, spaghetti).
(b) Shepherd's pie (meat, potato, plus milk, margarine).
(c) Moussaka (minced beef or lamb, tomato, cheese-sauce topping).
(d) Hamburgers with cheese (minced beef, breadcrumbs, egg, cheese) served with a garnish of red and green peppers.

BALANCED MENUS BASED ON MEAT

(a) Roast beef, Yorkshire pudding, potatoes, cabbage.
(b) Roast chicken, bacon rolls, bread sauce, new potatoes.

(c) Beef and ham pie, tomatoes; ice-cream or cheese and biscuits.
(d) Steak and kidney pie, greens, potatoes; caramel custard.
(e) Pork pie, coleslaw, rolls and butter; ice-cream, hot chocolate sauce.

STORAGE

Raw meat is best stored unwrapped, or wrapped in thin muslin, on a rack or hook in a cool airy place or cold store. Deep-frozen meat can be cooked unthawed if the piece is not too large.

Cooked meat is a potential source of food poisoning. If it has to be stored it should be cooled quickly and stored in a refrigerator. If it is re-heated, it should be *thoroughly heated* (i.e. 170 °F—77 °C) not merely warmed, since warming will speed up the growth of food-poisoning bacteria while thorough heating will kill them off.

OFFALS

CHIEF VALUE

Offals are of higher nutritional value than meat. They are equal to meat in protein and nicotinic acid content, richer in vitamin A (especially liver), thiamin, riboflavin and iron, and poorer in energy and fat. They are deficient in vitamin C and calcium.

FOOD PREPARATION

Overheating (i.e. above 170 °F—77 °C) makes liver and kidney granular and unappetizing. Quick cooking of the thinly sliced offals, by frying or grilling, keeps them tender and preserves the thiamin. Long slow methods, such as braising and stewing, produce tender savoury results, but reduce thiamin value; vitamin A and iron are unaffected.

BALANCED MENUS

These may be as given for meat; or stuff the liver or heart with cereal, bacon, herbs, etc. Offals can be made into mixtures with cereal and herbs, and steamed or baked, e.g. faggots, haggis.

STORAGE

Offals do not keep well, unless deep frozen. They should be kept cold and used as soon as possible after thawing.

TABLE 20

Offals—liver, kidney, heart—average composition per oz (28 g)

	kJ	*kcal*	*protein (g)*	*fat (g)*	*iron (mg)*	B_1 *(mg)*	B_2 *(mg)*	*Niacin (mg)*
Liver	160	40	6	1·5	3	0·1	0·7	4
Kidney	140	35	5	2	3	0·1	0·6	2
Heart	140	35	5	2	1·5	0·2	0·3	2

Vitamin A (μg R.E.) per oz liver; calf 4 150 μg (14 000 i.u.); ox 4 760 μg (16 000 i.u.); pig 2 000 μg (8 700 i.u.); lamb 5 000 μg (17 000 i.u.)

Average portion served 100 g (3½ oz)—no waste.

TABLE 21

Food value of average portion

A per portion; *B* expressed as fraction of adult allowance

		kJ	*kcal*	*protein (g)*	*iron (mg)*	B_1 *(mg)*	B_2 *(mg)*	*Niacin (mg)*	*vitamin A (μg R.E.)*
Ox liver	*A*	480	120	18	8	0·3	2·0	11	5 000
	B		1/20	1/3	2/3	1/4	100%	1/2	6 days
Kidney	*A*	430	110	14	8	0·3	1·8	5	240
	B		1/20	1/4	2/3	1/4	100%	1/4	1/3
Heart	*A*	430	110	14	4	0·5	0·8	5	50
	B		1/20	1/4	1/3	1/2	1/2	1/4	—

FISH

WHITE FISH

(Cod, haddock, coley) 30 to 40% waste.

Average portion served = 5 oz (140 g) = 9 oz (260 g) weighed with waste.

FOOD VALUE

White fish is a valuable source of good quality protein but is deficient in vitamins A, B_1, B_2, C and D, iron and calcium.

TABLE 22

Food value of white fish

	kJ	*kcal*	*Protein (g)*	*Fe (mg)*	*Ca (mg)*	B_1 *(mg)*	B_2 *(mg)*	*Nic. (mg)*
Content per oz	11	27	5	0·3	15	0·01	0·01	1·0
Proportion of daily allowance per 5 oz (140 g) portion		1/20	1/2	1/8	1/7	1/20	1/30	¼

BALANCED RECIPES

The deficiencies can be rectified by making fish cakes or fish pie (using white sauce, cheese, peas, hard-boiled eggs, with a potato or pastry topping, garnished with tomatoes). Pastry made with margarine supplies fat, carbohydrate, vitamin B_1, nicotinic acid, A and D. Calcium can be added with cheese and milk, iron with egg yolks, curry and shellfish (see also recipes for fish-cakes and fish chowder p. 126). Examples of balanced meals based on white fish are:

(a) Fish, chips, peas, orange in jelly, cheese and biscuits.

(b) Baked fish with onions, tomato, rice, green salad and fruit savarin.

FAT FISH

(Herrings, mackerel, salmon) 20 to 40% waste.

Average portion served = 4 oz (114 g) = 7 oz (200 g) weighed with waste.

TABLE 23
Food value of fatty fish

	kJ	*kcal*	*Fat (g)*	*Protein (g)*	*Fe (mg)*	*Ca (mg)*	*A (R.E. μg)*	*A (i.u.)*	*B₁ (mg)*
Content per oz	150	47	3	5	0·4	9	9–26	30–90	0·01
Proportion of daily allowance per 4 oz (114 g) portion		1/13	—	2/5	1/8	1/15	—	—	1/25

	B_1 (mg)	*Nic. (mg)*	*D (i.u.)*
Content per oz (28 g)	0·06	0·9	50–150
Proportion of daily allowance per 4 oz (114 g) portion	—	¼	50–100%

FOOD VALUE

Fat fish have greater nutritional value than white fish. They supply vitamin D, protein and nicotinic acid but little calcium, vitamins A, B_1, B_2 and C.

BALANCED RECIPES

The deficiencies can be rectified by serving with rice, potatoes, bread and flour (to supply carbohydrate), with eggs, cheese and carotene-rich vegetables (vitamin A), watercress, tomatoes, green peppers (vitamin C), cheese and milk (calcium) and vitamin B_2.

Kedgeree with rice, peas, tomatoes, green peppers and hard-boiled eggs with butter balances many of the deficiencies of the fish. Balanced meals based on fat fish include: soused herrings or smoked salmon with bread and butter, lemon and watercress; baked mackerel with bacon, onions, herbs and tomatoes with potatoes; fish-cakes with canned salmon, mixed with potato, butter and milk, served with bacon rolls, tomatoes and chips.

SHELLFISH

Molluscs (cockles, mussels, oysters) 75% waste. Crustaceans (lobsters, crabs, prawns) 63% waste.

Average portion served = 3½ oz (100 g) = 14 oz (400 g) weighed with waste.

TABLE 24

Food value of shellfish

Average composition: *A* per oz (28 g) of edible portion; *B*, fraction of daily allowance provided per average portion, 3½ oz (100 g)

		kJ	*kcal*	*Protein (g)*	*Ca (mg)*	*Fe (mg)*	*B_1 (mg)*	*B_2 (mg)*	*Nic. (mg)*
Molluscs	*A*	80	20	3	40	0·5–8	0·01	0·04	0·4
(shellfish)	*B*		—	1/5	1/4	10–200%	—	1/10	1/12
Crustaceans	*A*	100	25	5	30	1·5	0·01	0·06	0·7
	B		—	1/3	1/5	1/2	—	1/7	1/8

Shellfish provide good supplies of iron and protein and useful amounts of calcium, nicotinic acid and riboflavin. They are deficient in thiamin, vitamins C, A and D. For a balanced meal shellfish could be served with bread and butter, lemon and watercress; or prepared as seafood cocktail, with lemon and tomato juice, mayonnaise and toast; or a seafood pancake (yeasted) with a rich white sauce, served with green salad; or a paella (with rice, peppers, ham, tomatoes, oil, and saffron to flavour).

FOOD PREPARATION

Over-cooking makes fish dry and unappetizing because juices are lost when the protein hardens.

STORAGE

Fish very quickly loses its freshness and develops off-flavours, although this can be delayed by storing in a refrigerator or deep freeze. Flavour can be preserved to some extent by sprinkling with lemon juice or vitamin C crystals.

THE POTENTIAL IMPORTANCE OF FISH

Fish is a valuable food but is not eaten in quantity in many countries. While meat supplies a quarter of the protein, one-third of the nicotinic acid, one-fifth of the iron and riboflavin of the British diet, fish contributes only 5% of the total protein and 4% of nicotinic acid. Fish protein is similar to meat protein both in high quality and ability to supplement the lysine-deficient proteins of cereals.

Fig. 21 The difficulties of applying nutrition and promoting foods are world wide

We find the fish we want is too expensive: 146 families—48·7%

We do not like fish: 96 families—32%

We become ill when we eat fish: 21 families—7%

We find the fish is of bad quality: 7 families—2·3%

We do not find the fish we want in the shops when we want it: 2 families—0·7%

Other reasons: 28 families—9·3%

Of the world's total protein intake, fish contributes about 1%, but enough fish is available to increase this figure to 10%. Campaigns to eat more fish have taken place in many countries where fish is available (see Fig. 21). Fish farming is being developed, but progress is slow.

EGGS

Hen's eggs weigh approximately 2 oz (50 g) and the shell accounts for ¼ oz (7 g).

TABLE 25
Food value of eggs
Average composition per oz (28 g)

	kJ	*kcal*	*protein* (*g*)	*fat* (*g*)	*Ca* (*mg*)	*A* (*R.E.* μ*g*)	B_1 (*mg*)	B_2 (*mg*)	*Niacin* (*mg*)	*D* (μ*g*)
Whole egg	180	45	3·5	3·5	15	40	0·03	0·1	0·02	1
Yolk	400	100	4·5	9	60	115	0·09	0·1	0·01	3
White	45	10	2·5	trace	2	—	—	0·1	0·03	—

100 kcal portion is one large egg, i.e. 2¼ oz (64 g). An average portion (served as a main dish) is 4 oz (114 g) or two eggs.

FOOD VALUE

Two eggs supply the following proportions of the daily adult allowance: half of the vitamin A, riboflavin, protein, one-eighth of the thiamin and half of the vitamin D. Although egg contains iron it is not available.

Egg protein is 'top' quality (biological value 100) and is used as the reference protein standard for assessing protein quality of foods (see Chapter 4).

BALANCED MEALS

These include cheese omelettes, with tomatoes and wholemeal roll and butter; poached egg on chips and strawberry ice-cream; cheese savoury (eggs, milk, cheese, bread) with tomatoes; cheesecake, made with eggs, cottage cheese or condensed milk, butter, sugar, lemon in a pastry case.

OTHER PROPERTIES

Egg proteins hold air when beaten and coagulate when heated, they are therefore very useful to lighten doughs. Eggs can absorb and emulsify fats and are useful to bind mixtures. In particular egg yolk contains lecithin (a combination of phosphate, fatty acids, glycerol) which acts as an emulsifier of oil and water; egg yolk is therefore a valuable ingredient of salad creams and mayonnaise.

STORAGE

Eggs in shell keep well in the cold especially if the shell is sealed, e.g. by wax, waterglass or oil. Shelled eggs, both yolks and whites, can be stored deep frozen, and this is the product often used by bakers and confectioners. Dried eggs keep for many months, preferably stored in nitrogen-filled tins to prevent oxidative rancidity.

DUCK EGGS

Duck's eggs, unlike hen's eggs, are commonly infected with salmonella group bacteria and must be boiled for 10 to 15 minutes to ensure sterilization.

MILK

Cow's milk contains 88% water and 12% solids. Therefore 2½ oz (70 g) of dried milk makes up to 1 pint (20 oz or 560 ml) of liquid milk. The solids include 3·3% protein, 3·6% fat and 4·7% lactose (milk sugar). 'Channel Island milk' has 4·5% fat.
Goat's milk is similar to cow's milk in composition—3·3% protein, 4·5% fat and 4·4% lactose.
Sheep's milk is much richer in total solids—5·6% protein, 7·5% fat and 4·4% lactose.
Buffalo milk contains 3·8% protein, 7·5% fat and 4·9% lactose.

TABLE 26
Food value of cow's milk

	kJ	*kcal*	*Protein* (*g*)	*Ca* (*mg*)	*Vitamin A R.E.* *Winter* (μg)	*Summer* (μg)	B_1 (*mg*)	B_2 (*mg*)
Per oz (28 g)	80	19	0·9	35	8	11	0·01	0·04
Per pt (570 ml)	1 520	380	18	680	160	220	0·2	0·8
Proportion of daily allowance per pint		1/6	1/3	100%	1/5	1/3	1/5	1/2

These figures show that a daily pint of milk supplies a large proportion of the protein, calcium, riboflavin and vitamin A needed. A daily pint of skimmed milk, or the equivalent of dried, is equally good value for protein, calcium and riboflavin, and is much cheaper, but is deficient in fat and, therefore, vitamin A.

MANUFACTURED MILK PRODUCTS

There are various forms of milk generally available. (1) Full-cream dried milk, as the name implies, is simply whole milk evaporated to dryness. (2) Dried skim powder is fat free. (3) Half-cream dried milk is sometimes used for infant feeding. (4) Evaporated milk or condensed unsweetened milk, usually sold in cans, has been concentrated to about 31% solids, i.e. just over double the concentration of fresh milk. This is often fortified with added vitamin D, and, if it is intended for feeding babies, other vitamins may be added. (5) Condensed, sweetened milk may be skim or full-cream and has a considerable amount of sucrose added. As it is relatively poor in protein (expressed as a percentage of the total solids content) and as the skimmed product is free from fat and therefore free from vitamin A, sweetened condensed milk is not a suitable food for babies.

INFANT FEEDING

The product commonly used for infant feeding is dried full-cream milk. This is more expensive than the dried skim product which is lower in calorie content and lacks the vitamin A. It is possible to use skim milk powder for infant feeding by restoring the fat and vitamin A. Red palm oil, a rich source of carotene, or margarine fortified with vitamins A and D, is rubbed into the dried powder until it is no longer visible. When the milk is subsequently mixed with cold water the fat stays in suspension. This is not a recommended procedure in countries where the full-cream product is available but is useful when only skim milk powder is available.

TABLE 27
Composition of dried and condensed milks

A1—per oz (28 g) dry material or concentrated liquid; *A2*—per pint (570 ml) when diluted to be equivalent to fresh milk; *B*—fraction of daily adult allowance supplied by one pint (570 ml) of reconstituted milk

		kJ	*kcal*	*Protein* (*g*)	*Ca* (*mg*)	*A average* (*μg*) *R.E.*)	B_1 (*mg*)	B_2 (*mg*)
Skim powder	*A1*	420	100	10	360	trace	0·13	0·44
+ 7 parts water	*A2*	1 000	250	25	900	trace	0·33	1·1
	B		1/10	1/2	200%	—	1/3	2/3
Full cream powder	*A1*	560	140	7	250	90	0·09	0·33
+ 7 parts water	*A2*	1 400	350	18	600	225	0·22	0·82
	B		1/7	1/3	100%	1/3	1/5	1/2
Full cream condensed, unsweetened (31% solids)								
	A1	160	40	2	75	26	0·22	0·09
+ 1 part water	*A2*	1 600	400	20	750	260	0·2	0·9
	B		1/6	2/5	150%	1/3	1/5	1/2
Skimmed condensed, sweetened (71% solids)								
	A1	320	80	3	100	—	0·03	0·11
+ 5 parts water	*A2*	880	270	10	350	—	0·10	0·37
	B		1/10	1/5	3/5	—	1/10	1/4
Fresh whole milk	*B*		1/7	1/3	100%	1/4	1/5	1/2

USES OF MILK

Table 27 shows the valuable contribution that milk can make to the diet. One third of a pint—a glassful—which is a reasonable portion for a caterer to use, provides one third the calcium, a sixth of the riboflavin and a tenth of the protein and one fifth the vitamin A of the adult daily allowance.

Milk can easily be incorporated into a variety of dishes including cream soups, milk sauces and ice-cream. Fresh and concentrated milks can be mixed with iron-rich foods (egg yolk, chocolate) and vitamin C-rich fruits in milk shakes, mousses, soufflés, 'fools', and lemon meringue pie. It is widely used in puddings and with cereals and the nutritional value of such dishes can often be improved by using more milk or enriching the liquid milk with dried milk.

Dried milk, whole or skimmed, makes a good base for icings and spreads, both sweet and savoury, and for sweets such as fudge and milk toffee. It improves the nutritional value when added to doughs and batters but not more than 2 oz (57 g) can be used per lb (460 g) of flour without affecting the texture. Milk has the effect of 'tightening' the gluten of the flour giving a closer texture.

The 'milk loaf', by law in Great Britain, contains 6% added milk solids. It is not only nutritionally richer in calcium and quantity of protein but the milk protein supplements the missing lysine of the wheat protein and improves its quality.

KEEPING PROPERTIES

Fresh milk will keep for only a few hours unless it is chilled, pasteurized milk will keep for a few days but must be kept cool, sterilized milk will keep for several weeks. The principal reason for pasteurizing milk, however, is to make sure that harmful bacteria from a possibly diseased cow or from infected milk do not cause infection in the consumer. The most recently introduced product is known as UHT or ultra-high temperature treated milk and this will keep for several months until opened.

Pasteurization is a mild heat treatment and causes the loss of about 20% of the vitamin C, 10% of the thiamin and a smaller amount of the vitamin B_{12}. Since milk is, in any case, a very minor contributor to the vitamin C content of our diet these losses are a small price to pay for safe milk. Sterilization involves more severe heat treatment and destroys half of the vitamin C and one-third of the thiamin and nearly the whole of the B_{12}. The UHT process of sterilization does no more damage than pasteurization.

At temperatures above 170 °F (77 °C) one of the milk proteins, lactalbumin, is precipitated as a white 'fur' on the sides of the pan but the casein is not affected. If milk is kept hot a protein skin forms on the surface and leads to a loss of trapped fat and calcium. Milk kept hot for several hours can lose a considerable amount of its total solids in these two ways.

Yoghurt has the same nutritive value as milk. It is prepared by souring milk with *Lactobacillus bulgaricus*; the acidity clots the proteins and helps to preserve the milk.

CHEESE

Cheese is made of milk curd, produced either from soured milk or milk curdled with rennet (see p. 89). The milk is separated into the precipitated curds and liquid whey; the whey contains lactose, water-soluble vitamins and some calcium. A gallon of milk produces 1 lb of cheese and the average composition of cheese is one third each of water, protein and fat, with only traces of carbohydrate and vitamin D (and no waste). Flavours and food values depend on the type of milk and process used, and bacteria and moulds present.

TABLE 27A
Composition of cheeses (per 100 g)

Name	*Water* %	*kJ*	*kcal*	*Protein* (*g*)	*Fat* (*g*)	*Ca* (*mg*)	*vit. A* (*μg*)	B_2 (*mg*)	*Niacin* (*mg*)
Camembert	47	1 250	300	23	23	150	240	0·8	1·2
Cheddar	37	1 680	400	25	34	810	420	0·5	0·1
Cottage	76	400	100	15	4	80	30	0·3	0·1
Danish blue	40	1 470	355	23	29	580	330	(–)	(–)
Edam	44	1 260	300	24	23	740	270	0·4	(–)
Gouda	41	1 360	390	26	30	720	(–)	(–)	(–)
Gruyere	22	1 940	465	37	33	1 080	390	0·4	0·1
Mysost*	12	1 970	470	11	29	320	330	(–)	(–)
Parmesan	28	1 700	410	35	30	1 220	360	(–)	(–)
Stilton	28	1 920	460	25	40	360	480	0·3	(–)

* Carbohydrates: cottage cheese 4·5%, Mysost 42%.
(–) figures not available.

Average portion of Cheddar cheese served, 50 g (2 oz) provides about $\frac{4}{5}$ calcium, $\frac{1}{3}$ vitamin A, $\frac{1}{5}$ protein, $\frac{1}{6}$ riboflavin, recommended for an average adult.

FOOD VALUE

All cheeses are valued as good sources of protein (except cream cheese), calcium, vitamin A and riboflavin.

BALANCED RECIPES

Balanced cheese dishes can be made by combining with cereals (supplying iron and thiamin), eggs, meat and shellfish (supplying iron), fat fish and margarine (supplying vitamin D), and fruits and vegetables (supplying vitamin C).

Examples: grilled cheese on buttered toast (Welsh rarebit), plus poached egg (Buck rarebit), or kipper fillets (Aberdeen rarebit), served with tomato juice, or watercress, etc. Pasta with meat sauce, tomato and cheese (Bolognaise), or with cheese sauce and shellfish (Vongole) and green salad. Cheese hardens if heated over 170°F so cheese dishes must be quickly browned under the grill. As with milk, acidity helps with absorption of calcium, hence pickles with cheese.

Storage. Cheese keeps best in a cool place, wrapped in foil or muslin.

CEREALS

When man was a hunter and collector he lived on meat, fish, nuts, eggs, roots and berries. The discovery of cereals about 10 000 years ago made man settle down. Moreover, cereal cultivation ensured steady supplies and allowed time for social development.

Cereals include maize, wheat, rye, rice, barley and oats. They can be eaten whole as porridges or ground into flours and used to make pancakes, breads and cakes.

TABLE 28

Cereals: composition per oz (28 g)

	kcal	*Protein (g)*	*Fat (g)*	*Carb. (g)*	*Ca (mg)*	*Fe (mg)*	*B_1 (mg)*	*B_2 (mg)*	*Nic. (mg)*
Barley, whole, dehusked	100	3·4	0·6	19	10	1·0	0·1	0·05	2·0
pearled	100	2·6	0·4	22	6	0·2	0·04	0·02	0·8
Maize, whole	105	2·9	1·3	20	3	0·7	0·09	0·03	0·6
60% extn.	100	2·3	0·4	22	2	0·6	0·01	0·01	0·17
Oats, dehusked	110	3·4	2·1	19	17	1·4	0·1	0·04	0·3
Rice, lightly milled	100	2·3	0·4	22	3	0·6	0·07	0·01	0·6
polished, white	100	2·0	0·1	23	1·4	0·3	0·02	0·01	0·3
Rye 85–90% extn.	100	2·3	0·4	22	7	1·0	0·07	0·02	0·3
Wheat, whole	100	3·3	0·6	20	9	1·0	0·1	0·02	1·4
Flour, white, 70% extn.	100	2·9	0·3	21	5	0·4	0·02	0·01	0·2
enriched	100	2·9	0·3	21	33	0·5	0·08	0·01	0·7
Wheat bread, white, enriched	70	2·2	0·4	15	26	0·5	0·05	0·01	0·5
Oatmeal porridge (1 part oatmeal 8 parts water)	12	0·4	0·2	2	2	0·15	0·01	—	—

Cereals contain 10 to 15% water, 70 to 80% starch, 7 to 10% protein, with traces of fat, a positive amount of B vitamins (i.e. adequate to metabolize the starch), minerals, plus indigestible outer skins (bran). When cereals are made into flour the germ and bran can be removed during milling, and are described as offals. Because these offals average 30% of the whole grain, only 70% appears as white flour; this would be described as flour of 70% extraction rate. Wholemeal flours of wheat, rye and maize, containing all or almost all of the offals, are described as of 100%.

The average energy value of dry cereals, both whole and as flours is 100 kcal/42 kJ per oz, or 350 kcal/146 kJ per 100 g.

Whole grain cereals provide useful amounts of iron and B vitamins. Highly-milled cereals (i.e. low extraction rate) do not provide useful amounts of these nutrients unless they are enriched. Because cereals are usually eaten in rather large amounts they are an important source of protein.

TABLE 29
Food value of average portion—7 oz (200 g)
A per portion; *B* expressed as fraction of adult daily allowance

		kJ	*kcal*	*Protein (g)*	*Ca (mg)*	*Fe (mg)*	B_1 *(mg)*	B_2 *(mg)*	*Niacin. (mg)*
White bread, enriched	*A*	200	480	15	180	3·5	0·35	0·07	3·5
	B		1/5	1/3	1/3	1/4	1/3	—	1/6
White flour, enriched	*A*	300	700	21	230	3·5	0·56	0·07	3·5
	B		1/4	2/5	1/2	1/4	1/2	—	1/6
White rice (1 part rice, 2 parts water)	*A*	100	230	4	2	0·2	0·05	0·02	0·9
	B		1/10	1/12	—	—	1/20	—	1/20
Oatmeal porridge (1 part oatmeal, 8 parts water)	*A*	40	90	2·6	13	1·1	0·08	0·03	0·2
	B		1/30	1/20	—	1/10	1/12	—	—

BALANCED RECIPES

If cereals are served with milk or cheese and green vegetables or vitamin C-rich fruits, and small amounts of such animal protein foods as meats, fish or eggs, then balanced meals result. In many countries where cereals form a large part of the diet, mixtures of this kind have become the national dish, e.g. couscous in North Africa, paella in Spain, Jollof rice in Nigeria, risotto in Italy, pillaf in Eastern Europe.

Calcium can also be added to cereals during preparation, as is done in Mexico where maize is soaked in lime water before use.

(a) *Rice*. Risotto (tomato, fat, cheese); *paella (oil, shellfish, meat, peppers, tomatoes); *Jollof rice (oil, shellfish, meat, green leaves, nuts); kedgeree (rice, fat, fish, eggs, curry).

(b) *Oats*. Muesli (oats, lemon, nuts, milk, apple); white puddings (oatmeal, onions, fat) served with bacon, tomatoes and milk; haggis (liver, oats, onions) served with swedes.

(c) *Wheat*. Couscous (steamed whole wheat with meat stew, beans, carrots, peppers); cheese scones served with watercress; Cornish pasties (pastry with meat, potato, onion and parsley).

* Appendix, p. 193.

(d) *Maize.* Tortillas (maizemeal pancakes served with filling of stewed meat and tomatoes, peppers).

N.B. Cornflour (cornstarch in U.S.A.) is prepared by grinding maize with water and separating the starch from the protein. It consists of almost pure starch with only traces of other nutrients; it is valued for its thickening qualities in cooking.

FIBRE IN DIET

The indigestible bran or fibre has been claimed to have a beneficial effect on the muscles of the large and small intestines and its absence is said to cause various disorders. The use of some whole grain cereals or the addition of a small amount of bran (1–2 oz daily) to breakfast cereals and doughs has therefore been recommended as a useful health measure.

STORAGE

Cereals are liable to attack by insects and pests and should be stored in closed containers in a cool place. The containers and bins should be made of metal or plastic, with rounded corners and straight edges so that pests cannot lodge. Large bins should be mounted on castors, so that the cereals can be used in the kitchen and returned to store in a cool place.

BREAD

Bread is made from a risen dough of wheat flour, yeast, salt and water. The gluten in the dough is elastic and is blown up into 'balloons' by the expansion of gas from the yeast fermentation, then set by heat giving a light, spongy texture. To make a light, risen loaf, it is essential to use a cereal containing gluten, such as wheat or rye. A variety of breads can be made by mixing in other cereals with the wheat flour.

Because bread contains 37% water, compared with 10% in the flour, the nutritional content of bread is approximately 75% of that of flour used.

Brown breads can be made from wholemeal (100% of the grain, i.e. 100% extraction rate), wheatmeal (85 to 95%) or from speciality flours, which are mixtures of white flours with extra amounts of wheatgerm.

White bread is usually 72% extraction, i.e. 28% of the grain is removed as bran and germ; special loaves have added protein.

Food value All breads, white and brown, are valued as sources of energy and protein, and brown for iron and B vitamins. In many countries the white flour (and consequently the white bread) is fortified with vitamins B_1, B_2 and nicotinic acid (B_2 is not used in Great Britain) together with iron, so that it is equivalent in nutritive value to bread of 80% extraction (and somewhat inferior in B vitamins and iron to wholemeal bread). Calcium is added to provide larger amounts than are found in wholemeal bread. (See p. 156.)

In Great Britain bread, flour and cereals supply nearly one-third of the average energy intake, nearly one-quarter of the protein, a quarter of the calcium and one-third of the vitamin B and nicotinic acid.

Staling of bread Staling occurs when water passes from the gluten to the starch particles, and takes place more rapidly at temperatures of 35 to 65 °F (2 to 18 °C), i.e. in a cool larder or refrigerator. Bread should be cooled slowly after baking and will keep reasonably fresh for 3 to 4 days at room temperature stored in a perforated polythene bag or bread bin fitted with a loose lid.

Bread will keep indefinitely in the deep freeze. It should be thawed out at room temperature for half an hour or in a hot oven for 5 minutes.

The staling effect can be partly reversed, i.e. bread can be refreshed, by heating in a closed tin in a hot oven for 10 minutes, allowing it to cool in the closed tin.

CONCLUSION

Cereals are popularly regarded as starchy foods but even in the quantities eaten in Great Britain they contribute as much protein to the average diet as the acknowledged protein foods such as meat and milk.

In the developing countries where cereals may constitute up to 80% of the total intake they are often the major source of protein in the diet.

NUTS AND SEEDS

Nuts are dried seeds, and sesame and sunflower seeds should be included in this group. The protein content is high but of moderate quality and needs to be supplemented with animal protein, or mixed with cereals and pulses to provide the right proportions of essential amino acids (see p. 60). Peanuts are pulses, i.e. leguminous plants, but their composition is much closer to that of nuts than of pulses. (See Table 30.)

TABLE 30
Composition of nuts and seeds

A per oz (28 g);
B fraction of daily adult allowance in average portion of 2 oz (57 g)

		kcal	*Protein (g)*	*Fat (g)*	*Carb. (g)*	*Ca (mg)*	*Fe (mg)*	B_1 *(mg)*	B_2 *(mg)*	*Nic. (mg)*
Almonds	*A*	190	6	17	3·5	40	1	0·1	0·2	1·3
	B	1/6	1/4			1/6	1/6	1/5	1/4	1/7
Brazil	*A*	200	4	19	2·6	50	0·9	0·3	0·03	2·3
	B	1/6	1/6			1/5	1/6	1/2	—	1/4
Cashew	*A*	170	6	13	7	14	1·4	0·2	0·06	0·6
	B	1/7	1/4			—	1/5	1/2	1/10	1/15
Sesame seeds	*A*	170	6	14	4·6	430	2·8	0·3	0·07	1·4
	B	1/7	1/4			100%	1/2	1/2	1/10	1/7
Peanuts (dried)	*A*	170	8	14	5	17	0·7	0·26	0·04	4·8
	B	1/7	1/3			—	1/10	1/2	—	1/2
Sunflower seeds	*A*	150	8	10	6·5	30	2·0	0·5	0·06	1·7
	B	1/8	1/3			1/8	1/3	100%	1/10	1/5
Coconut (desiccated)	*A*	180	2	17	1·8	6	1·0	0·01	0·01	0·1
	B	1/7	1/12			—	1/6	—	—	—

FOOD VALUE

Nuts are valuable for energy, protein, iron and B vitamins, except for chestnuts and coconuts, which are much poorer in these nutrients. Nuts lack vitamins A, C and D and most of them are low in calcium (except almonds) and in carbohydrate, but this makes them a useful food for diabetics and slimmers. Nuts are used whole, ground as meal, or as creams or pastes (e.g. peanut butter).

DIGESTIBILITY AND STORAGE

Nuts are difficult to digest unless grated or pounded. If used as a main dish, like pulses, they may lack the savoury taste which stimulates the appetite, but this can be provided with yeast extracts, monosodium glutamate, dried onions, herbs, etc. They are best stored whole or as pastes. When milled they do not keep well and can become rancid.

BALANCED RECIPES

Vegetarian mixtures of good food value contain nuts, dried milk, soya and cereals, flavoured with savoury extracts, dried herbs, mushrooms, etc. Such mixtures can be used to make rissoles, baked 'roasts', and as stuffing for peppers, tomatoes and baked onions. Nut pastes can be made into balanced snacks with bread, served with salad and milk; they can also be made into drinks and creams, mixed with fruit juices, eggs, milk and sugar. Milled and pounded nuts can be made into soups and spreads and milks.

(a) Nut breads; sweet-walnut and apricot bread, almond and currant bread, savoury-peanut bread with brewers' yeast.

(b) Ground nut stew, Nigeria; a savoury mixture of stewed meat, smoked dried fish, peppers, tomatoes, and pounded peanuts.

(c) Chicken with almonds (Chinese dish); steamed chicken in savoury sauce, with salted almonds.

(d) Chicken almond soup; chicken stock, ground almonds, egg yolk, cream and rice.

(e) Savoury nut spread; nut paste, soya, potato, onion, yeast extract.

(f) *Nut fudge (uncooked); pounded peanuts, honey, dried skim milk.

* See Appendix, p. 197.

FRUITS AND VEGETABLES

Most fruits and vegetables average 85 to 90% water, 1 to 2% protein and 2 to 4% carbohydrate as starch or sugar. The rest of the solid matter is indigestible cellulose which does, however, serve a useful purpose in providing bulk and stimulating the intestinal muscles so preventing constipation.

Almost all vegetables and fruits are poor sources of protein, fat, iron and calcium. The thiamin levels appear to be low, but because of the low energy value, especially of green leaves, the ratio of thiamin to energy is high (i.e. more than 0·4 mg per 1 000 kcal) and so fruits and vegetables make a positive contribution of vitamin B_1 to the diet. Where the staple food is poor in thiamin, for example white rice or cassava, green vegetables are a valuable addition to the diet.

Vitamin C rich fruits and vegetables—peppers (sweet, green and red), black and red currants, strawberries, paw-paw, cabbage, broccoli, sprouts, cauliflower, radish, kale, citrus fruits, gooseberries, tomatoes, watercress, spinach, okra, and spring onions.

Poor in vitamin C—celery, onions, aubergines, salsify, plums, cherries, pears, apples, peaches, apricots, lettuce, cucumber, marrow, parsnips, carrots, beetroots, artichokes and bamboo shoots.

Carotene—all green vegetables contain carotene; carrots are an exceptionally rich source, and yellow fruits like apricots and mangoes are a moderate source.

BALANCED RECIPES BASED ON VEGETABLES AND FRUITS

To make balanced recipes and meals where large amounts of fruit and vegetables are eaten, protein, calcium and iron-rich foods must be added.

(a) Vegetable hot pot; carrots, potatoes, onions, celery, tomatoes, served in a milk sauce, with grated cheese and enriched with egg.

(b) Vegetable salad; lettuce, watercress, tomatoes, potato salad (with mayonnaise), hard-boiled egg or shrimps, cream cheese.

(c) Fruit salad; pears, peaches, oranges, in sugar syrup, served with rich ice-cream or custard.

STORAGE AND PROCESSING

Since vitamin C is the most unstable of nutrients, fruits and vegetables can lose the greater part of their food value in storage and preparation. When dried in hot air they can lose one-third or more of the vitamin C, but less if dried in vacuum. The new and gentler methods of processing, such as accelerated freeze-drying and deep-freezing cause much smaller losses, only about 10%. During the blanching process prior to drying or canning some vitamin C, together with other water-soluble vitamins, is washed out into the water and 5 to 20% may be lost in this way, depending on the size of the pieces and conditions of the blanching. After the initial processing loss, the vitamin C keeps well in dried, canned and frozen foods for periods of several months.

STARCHY ROOTS AND FRUITS

This group of foods includes potatoes (both Irish and sweet), yams, cocoyams (taro), plantains and cassava (or manioc root), and the flours prepared from arrowroot and sago palm. They are relatively easy to grow with a high yield and form the major part of the diet in many parts of Africa and the Pacific.

The description 'starchy' is well-merited as these foods are very low in protein content and are almost free from fat. Where any one of them is the staple food of a country there is almost always an accompanying shortage of protein in the diet.

TABLE 31
Composition of fruits and vegetables

A content per $3\frac{1}{2}$ oz (100 g) portion
B expressed as a fraction of daily adult allowance

		Ca (*mg*)	*Fe* (*mg*)	*A* (*µg R.E.*)	B_1 (*mg*)	B_2 (*mg*)	*Nic.* (*mg*)	*C* (*mg*)
Daily allowance		500	12	750	1·0	1·5	18	50
Cabbage, cooked	*A*	60	0·5	45	0·03	0·03	0·15	20
	B	1/8	—	—	—	—	—	2/5
Broccoli, cooked	*A*	160	1·5	420	0·06	0·2	0·6	40
	B	1/3	1/8	1/2	1/15	1/8	—	4/5
Carrots, old	*A*	40	0·4	2 000	0·05	0·04	0·4	4
	B	1/12	—	3 days	1/20	—	—	1/12
Tomatoes	*A*	13	0·4	15–120	0·06	0·04	0·6	20
	B	—	—	—	1/15	—	—	2/5
Peppers, red	*A*	20	1	80	0·06	0·08	1	150
	B	—	1/12	—	1/15	—	—	3 days
Orange	*A*	41	0·3	7	0·10	0·03	0·2	50
	B	1/12	—	—	1/10	—	—	100%
Gooseberries, stewed with sugar	*A*	20	0·25	35	—	0·02	0·2	28
	B	—	—	—	—	—	—	3/5

TABLE 32
Average composition per oz (28 g) of starchy roots and fruits

	kcal	*Protein* (*g*)	*Fe* (*mg*)	B_1 (*mg*)	B_1 (*mg*/1 000 *kcal*)	B_2 (*mg*)	*Nic.* (*mg*)	*C* (*mg*)
Cassava, fresh	45	0·2	0·3	0·02	**0·45**	0·01	0·2	10
flour	100	0·4	0·6	0·01	**0·11**	0·01	0·2	0
Plantain	35	0·3	0·1	0·02	**0·4**	0·02	0·2	7
Potato, Irish	20	0·6	0·2	0·03	**1·3**	0·01	0·4	5
Potato, sweet	30	0·4	0·3	0·03	**0·87**	0·01	0·2	5
Taro	30	0·6	0·3	0·03	**0·9**	0·01	0·3	2
Yam, fresh	30	0·6	0·3	0·03	**0·95**	0·01	0·1	3
flour	90	1·0	2·8	0·04	**0·5**	0·03	0·3	0
Arrowroot flour	100	trace	0·3	trace	**0**	0	0	0
Sago flour	100	0·1	0·3	trace	**0**	0	0·07	0

The table of composition shows that they contain reasonable amounts of the B vitamins (except for arrowroot and sago flours) and bearing in mind that the allowance of vitamin B_1 is 0·4 mg per 1 000 non-fat kcal, the starchy roots and fruits are very good sources of this vitamin.

In Great Britain potatoes are the most important single source of vitamin C, providing about half of the total supply in winter time. An average portion of 7 oz (200 g) provides not only a quarter to a half of the daily allowance of ascorbic acid but also one-sixth of the thiamin and a twelfth of the iron. The ascorbic acid content of potatoes falls steadily during storage, from 30 mg per 100 g (9 mg/oz) when freshly dug, to about 8 mg (2 mg/oz) after 8 months' storage. About one-third of the vitamin is lost in cooking. Therefore supplies of vitamin C must be safeguarded during late winter and early spring when our major supply is failing and fruits and green vegetables are scarce and expensive.

BALANCED RECIPES

These starchy foods are so grossly deficient in protein that it is essential to balance them with protein-rich foods. For example:

(a) Chowder—a type of stew based on potatoes, yams or plantains with milk sauce, fish, pork, carrots and shellfish.

(b) Fish-cakes—mashed potato mixed with canned or cooked fish, milk, oil, egg.

(c) Savoury bake—potatoes baked with milk, cheese, margarine.

(d) Sweet potato pudding—grated potato baked with custard made from eggs, milk and sugar.

(e) Potato pancakes—grated raw potato mixed with flour, milk and eggs fried in oil.

(f) Hotpot—minced or sliced beef and kidney with sliced raw potatoes, milk, sage and onion, baked slowly and served with tomatoes.

(g) Fritters—plantain with a batter of flour, egg, yeast and milk, cooked in oil.

(h) Vegetable hotpot—stew of beans, plantain, green leaves, yeast extract, dried milk and egg.

PULSES

Pulses are seeds of the leguminosa family which consists of peas, beans and lentils. The mature seeds contain only 10 to 13% water and so are concentrated foods. Their protein content is higher than that of cereals at 20% of dry weight, and they are good sources of thiamin, niacin and iron.

Although they do not contain any vitamin C this vitamin is formed during germination, so pulses that have been soaked in water until they begin to sprout become a useful source (see Appendix p. 000).

In many parts of the world where the diet is based largely on low-protein starchy foods and fruits, pulses (including groundnuts) are the most important source of protein, they are sometimes called 'poor man's meat'. Even in the British diet where the generally accepted protein foods such as meat, cheese, milk, etc., are freely available, pulses are the cheapest source of protein.

TABLE 33
Average composition per oz (28 g) of pulses

	kcal	*Protein* (*g*)	*Fat* (*g*)	*Ca* (*mg*)	*Fe* (*mg*)	B_1 (*mg*)	B_2 (*mg*)	*Nic.* (*mg*)
Chickpea	100	6	1·5	35	2·5	0·15	0·04	0·4
Cowpea	95	6	0·4	25	1·5	0·25	0·04	0·5
Groundnuts (dried)	160	8	12	15	0·7	0·25	0·04	5·0
Kidney bean	95	7	0·5	30	2·2	0·15	0·06	0·6
Lentil	95	7	0·3	20	2	0·15	0·06	0·6
Butter bean	95	6	0·4	30	1·7	0·15	0·04	0·4
Pea (dried)	95	7	0·3	20	1·5	0·22	0·06	0·7
Soya beans	110	10	5·0	60	2·0	0·3	0·09	0·6
flour, full fat	120	11·5	6·7	60	2·0	0·2	0·08	0·6
flour, low fat	95	14	2·0	70	2·6	0·25	—	0·65
'milk'	10	1·0	0·4	—	0·2	0·03	0·01	0·06
curd	20	2·0	1·1	10	0·5	0·01	0·01	0·1
sauce	—	1·7	0·3	30	1·6	—	0·04	0·1

BALANCED RECIPES

Mixtures of pulses and cereals provide good quality protein. This is because cereal proteins are rich in methionine and pulse protein in lysine. To make balanced recipes or meals, add calcium with enriched flour or bread, or milk, and vitamin A with milk, margarine or butter, plus vitamin C-rich fruits and vegetables.

1. Baked beans on toast (enriched bread with margarine) and tomatoes or tomato juice, or watercress.
2. Lentil soup, made with tinned tomatoes, served with croutons fried in margarine.
3. Mexican Bean Pot. Soaked red beans cooked with carrots, tomatoes (vitamin A), onions, garlic, chillis and oil, served with pancakes of maize meal (mixed with skimmed milk) and chopped red peppers.

4. Bean salad. Hot cooked red beans and white beans, mixed with oil, mustard and vinegar. When cold stir in chopped raw onion, parsley, peppers, and cubes of cheese, boiled rice.
5. Lentil curry with rice.
6. Dahl Pouri.* Yeasted pancakes of flour, skimmed milk and yeast filled with lentil curry or baked beans and served with chopped peppers.

NAMES

Much confusion has been caused by the different names that are used for the pulses in different countries. The following is a short list:
(a) Lima bean, butter, curry, Madagascar, sugar—*Phaseolus lunatus.*
(b) Navy bean, pinto, snap, haricot, kidney, French (unripe)—*Phaseolus vulgaris.*
(c) Kaffir pea, cow, black-eyed—*Vigna* species.
(d) Broad bean, horse, fava—*Vicia faba.*
(e) Mung bean, black gram, woolly pyrol—*Phaseolus mungo.*
(f) Field bean, hyacinth, Egyptian kidney, tonga—*Dolichos lablab.*

Dahl is the Indian name for split peas, e.g. lentil is red dahl, pigeon pea is red or yellow dahl.

Gram is the Indian name for small dried peas, e.g. pigeon pea is also called red gram, mung bean is black gram, chickpea is Bengal gram.

SOYA BEANS

Soya beans play a particularly important part in many national diets. They contain more protein (35%) of higher biological value (70) and more fat (18%) than the other pulses. The sprouted beans contain 4 mg vitamin C per oz.

The powdered bean is bitter to taste and contains toxins (including an anti-trypsin factor that interferes with the digestive enzyme) but both of these factors are removed by heating. The soya flours commercially available have all been subjected to heat treatment.

There are many traditional foods made from soya including fermented soy sauce, rather like yeast extract in flavour, other fermented products such as tempeh and natto, soya milk (a water extract of the bean) and soya curd (precipitated from the 'milk'). In the industrialized countries the soya bean is a major source of oil and protein supplement. The full-fat soya of commerce is simply the powdered, dehusked bean after a preliminary heat treatment; it is also available as the fat-free powder of around 50% protein content. (See also p. 71.)

Cooking properties Soya contains lecithin which has emulsifying properties. It does not contain starch and therefore does not thicken on heating.

Soya flour is widely used in sausages, meat and fish pastes, as a bread improver, and in biscuits, cakes, pastas and soups. Its high protein content makes it a valuable supplement to such foods. Its low carbohydrate content allows it to be used in slimming and diabetic foods.

Recipes Wheat flour enriched with 30% of soya flour will produce a loaf containing 20% protein compared with the normal level of 10%. Such enriched bread has more good quality protein than an equal weight of meat and is a valuable method of adding protein cheaply to the diet.

(a) Enriched bread, 1·1 lb (450 g) flour, 2 oz (57 g) soya flour, 1 oz (28 g) yeast, salt, water, margarine (protein approximately 3 g per oz (10 g per 100 g) compared with the normal level of 2·2 g (7·7 g per 100 g)).

(b) Enriched bread, 2·1 lb (450 g) flour, 4 oz (114 g) soya flour, yeast, salt, water, egg.

(c) Noodles, macaroni—1 lb (450 g) flour, 2 oz (57 g) soya flour, salt, water.

(d) Marzipan—6 oz (170 g) soya flour stirred into 3 oz (85 g) soft margarine in ¼ pint (140 ml) cold water; add 12 oz (340 g) sugar plus flavouring.

(e) Sausages—9 lb (4 kg) lean beef, 3 lb (1·4 kg) fat, 8 oz (230 g) soya flour, 1 pint (570 ml) water plus seasoning.

(f) Bean curd—fried in oil served with vegetable stew.

(g) Savoury spread—dried potato powder, soya flour, dried onion, yeast extract, herbs, butter, mixed with boiling water.

SUGARS, SYRUPS, JAMS

This group of foods consists almost solely of carbohydrate with no other nutrients (except where specified below). Not only are they devoid of protein, minerals and fats but they do not contain enough thiamin to metabolize the carbohydrate that they contain. It is this absence of other nutrients that has given rise to the term 'empty calories' for these sugary foods.

All white sugars, i.e. granulated, caster, icing, consist of 100% sucrose and nothing else. They differ from each other only in particle size. The brown sugars contain traces of other nutrients but apart from iron these are too small to be significant in the diet. Crude brown sugar contains 1% water, a trace of protein, and in quantities per oz—

8 mg calcium, 0·6 mg iron, 0·006 mg thiamin, 0·03 mg riboflavin and 0·1 mg nicotinic acid. It must be borne in mind that many of the brown sugars on the market are not this crude brown sugar but are various stages in the refinement of the sugar and so they contain even less of these other nutrients.

The raw molasses from which the brown sugar is made does contain significant amounts of iron and calcium. The quantities per oz are 45 mg calcium and 1·1 mg iron in light molasses and 170 mg calcium and 9 mg iron in the dark molasses. Thiamin content is 0·015 mg, riboflavin 0·015 mg and the nicotinic acid is 0·4 mg per oz.

Syrups are either refined molasses or solutions of pure sugar and in both instances can be regarded as pure sucrose.

It is sometimes believed that honey, because it is consumed in its 'natural' state, has some extra nutritive value but this is not true. It contains only minute traces of calcium, iron and thiamin, and only small amounts of riboflavin—0·015 mg per oz—and nicotinic acid—0·06 mg per oz. The carbohydrate of honey is not sucrose but is a mixture of glucose and fructose, known as the invert sugars.

Although jam is sugar-preserved fruit it has virtually no nutritional value apart from the energy content of the sugar. In order to prevent growth of moulds and yeasts jam must contain not less than 65% soluble solids. The small amount of fruit present makes no nutritional contribution apart from the vitamin C. This ranges between zero and 12 mg per oz, depending on the fruit and the method of making the jam. For example boiling the jam in copper vessels would completely destroy the vitamin C. However, even under the best conditions the contribution of jam towards our vitamin C intake is very small so such losses have no nutritional significance.

In general the sugars, syrups, jams and honey should not be considered as sources of any nutrients, only energy. For most people they improve the palatability of foods but this, in itself, may sometimes be undesirable as it can lead to over-eating. Moreover there is considerable evidence to show that a high sugar consumption is one of the factors involved in heart disease. So the rule should be—look on sugars as a 'fun food' but do not overdo it.

EXAMINATION QUESTIONS

1. What contribution is made by fruits and vegetables to the nutritive value of human diets? Which fruits and vegetables are good sources of the important nutrients? (R.S.H.)
2. Describe how you would use 'left-overs' of:

boiled potatoes	bread and butter
carrots	soured milk and cream
cabbage	fried fish
chips	shepherd's pie

What would then be their nutritional value?

3. Why is milk considered to be an ideal food for children? Suggest four ways of serving milk to a three-year-old child who dislikes it.
4. What types of milk are available for large scale cooking? Compare their nutritive values. Explain how you would use them, and give examples of dishes in which each may be used. (R.S.H.)
5. What precautions would you take during storage to retain the nutritive values of meat, milk, vegetables and fruit? (R.S.H.)
6. Describe in detail how to make one suitable accompaniment for each of the following, giving reasons for your choice: (a) steamed white fish, (b) a green salad, (c) a fruit pie, (d) hard-boiled eggs.
7. Compare the nutritional values of present-day white and wholemeal bread. Which would you advise for children at a boarding school, and why? (R.S.H.)
8. What is the food value of an average portion of white fish? Plan two main meals based on fish, using other foods to supply nutrients which are absent or in poor supply. Why and how would you promote the increased use of fish in a popular restaurant, and in a high-class snack bar? (Hotel Management Diploma.)
9. You are asked to economize in catering for a hospital by: (a) using margarine instead of butter, (b) less fresh milk and more dried milk, (c) more apples and home grown fruit instead of oranges, (d) more bread and less cakes and biscuits. Give your observations on this new policy. Indicate how you could adapt to these changes, and still maintain good food value. (Hotel Management Diploma, also R.S.H.)
10. How would you rate liver, milk and lemons as articles of diet? Answer this question under the following headings: (a) value for money, (b) nutritional value, (c) general use in cooking. (City & Guilds Adv. Dom. Cookery.)
11. Plan suitable recipes on behalf of Carnation Milk for the 'Family Circle' magazine, giving reasons for your choice. (Isleworth Diploma in Home Economics.)
12. Prepare a bread dough using 1·5 kg flour and use it to make rolls, a loaf, pizza and a yeast cake. Using pizza as the main course, prepare a balanced 3-course meal for two college students (men). What foods could you have added to the dough to increase the

iron and thiamin value of the bread? Calculate the energy value of a portion of the yeast cake. ('A' level practical, Home Economics.)

13. The magazine *Which* compared the weights of various types of meats supplying the same amounts of protein and food energy as 1 lb topside of beef.

Meat	*Wt to get same energy* (lb)	*Wt to get same protein* (lb)	*cost/lb*
Beef. Topside	1	1	55p
Grilling steak	1	1	80p
Stewing steak	1	1	45p
Shin of beef	$1\frac{1}{2}$	1	40p
Lamb. Loin chops	1	$1\frac{1}{2}$	30p
Shoulder and bone	$1\frac{1}{4}$	$1\frac{1}{2}$	25p
Pork. Loin chops	$\frac{3}{4}$	$1\frac{1}{2}$	40p
Lean belly pork	$\frac{1}{2}$	$1\frac{3}{4}$	20p

Using the given prices, calculate which are the 'best buys' for (a) an active man, (b) a teenage boy immobilized with a broken leg. What nutrients are poorly supplied by meat? Prepare and serve two or more dishes using the 'best buys' for (a) and (b). Add suitable foods and accompaniments to make a balanced main meal.

Arrange your work with appropriate notices as a display for an Open Day at school. Calculate the energy value of one dish. ('A' level practical Home Economics.)

14. When shopping for food it is essential to know the nutritional value obtained in relation to the money spent. Compare two foods in each of the following pairs for nutritive value in relation to cost.
(a) Sliced white bread and wholemeal bread
(b) Milk and cider
(c) Kippers and halibut
(d) Pork fillet and pig's liver
(e) Cheddar cheese and Camembert cheese
Discuss other factors that influence food prices. (London 'A' Level 1976.)

15. Compare the nutritional values of fresh, frozen, and dried vegetables. What methods of cooking would you recommend for the following: peppers, celery, beetroot, aubergines, and asparagus?

8 Manufactured foods

Very few foods come to us in the raw state. Only fresh fruits and vegetables, meat, eggs and fresh fish are untreated—all other foods have been processed in some way.

There are several reasons for this, the most important being the need to preserve food—by methods such as pasteurization, sterilization, drying, freezing, smoking, salting and pickling. A second reason is to improve the palatability of raw foodstuffs which may not be very attractive in their original state, such as wheat or oilseeds.

A third reason for processing is to add variety, for example a whole range of breads, cakes, biscuits, puddings, pastas and breakfast cereals are manufactured from a few basic ingredients. Preservation may also add to variety since processes such as pickling and smoking produce foods often quite different from the original, e.g. compare smoked salmon, kippers or bloaters with the fresh form.

Many of these processes, both those intended for preservation and to produce variety, are carried out by the housewife in her own kitchen. Nowadays, however, much of it is done for her in the factory. Not only may she buy ready-made bread, cakes, meat pies and canned puddings, but also 'instant' foods which need only the addition of water before being eaten. In addition the manufacturer can do some things that even the most skilled housewife or chef cannot do, he can freeze fish at sea immediately after catching so that it remains fresh, he can dry foods under controlled conditions, and he can exercise precise control over food preparation by means of elaborate laboratory techniques.

Yet another reason for processing is that it is sometimes necessary to remove or destroy harmful substances naturally present. For example cassava contains prussic acid in sufficient amounts to be poisonous, and it has to be treated before it can be eaten. Soya and many other beans contain substances that interfere with the digestive enzyme, trypsin, and they must be heated before they are safe to eat. Heat processing often causes a loss of nutrients but in some instances it improves nutritive value by increasing digestibility and by making some nutrients more available (see also phytic acid, pp. 38, 75).

The enrichment of foods by the addition of vitamins, minerals and protein is also part of food processing.

PRESERVATION

As soon as animals or fishes are killed, or plants harvested, they are attacked by bacteria, yeasts and moulds. The first effect of this is to change the flavour considerably, in most cases making the food quite unacceptable, e.g., bad meat or fish. In a few instances these effects are encouraged because they produce a flavour that is liked, for example Chinese eggs, soya sauce and sour milk. A consequence of more importance than deterioration in flavour is that foods may become dangerous to health if the attacking bacteria are pathogenic (disease-producing).

There are two further ways in which foods can deteriorate during storage. They can be spoiled by the action of enzymes naturally present in the tissues, and they can be oxidized through exposure to the air, as for example, when butter goes rancid.

Food, therefore, cannot be kept for long unless these activities are stopped or at least slowed down. This is the purpose of preservation.

Cooling The growth of microbes (bacteria, yeasts and moulds) is slowed down by cooling and stopped completely by deep-freezing, i.e., at temperatures as low as 0 °F (−18 °C). The microbes are not killed by cold but merely slowed down, so as soon as the temperatures rises spoilage sets in again. It is most important to realize that food heavily contaminated with bacteria before putting into the refrigerator is just as badly contaminated when it is taken out again.

Cooled food (32 to 40 °F, 0 to 5 °C) may keep well for several days. Frozen food (25 to 28 °F, −5 to −2 °C) may keep well for a few weeks. Deep frozen food (0 °F, −18 °C) will keep for several months, *but must not be allowed to thaw out and be refrozen*. It must be kept at this very low temperature at all times. Deep-frozen food taken from the commercial deep-freeze and put into the domestic refrigerator may still remain frozen hard but it is not cold enough unless it is put into the deep-freeze section.

Deep-frozen foods retain their nutritive value very well indeed for months at a time. There is some loss of vitamin C when vegetables are kept cool (32 to 40 °F, 0 to 5 °C), but this is very much less than the losses that take place at room temperature.

Heating Instead of slowing down the growth of microbes they may be killed off altogether by heat. If the food is heated sufficiently to destroy *all* the organisms it becomes sterile, and, if protected from recontamination, as when bottled or canned, will remain unchanged. **(A can of meat was opened 143 years after it had been canned and found to be sterile.)**

However, it is not always necessary to destroy all the organisms, but it may be enough to destroy only the disease-carrying ones (pathogens). This may be achieved by heating less severely—the process of pasteurization—and has the advantage of not producing as much of a cooked flavour as does the sterilization process. A comparison of the colour, flavour and keeping properties of pasteurized and sterilized milks illustrates this point. Pasteurization has no effect on the colour and little on the flavour but it lengthens the storage life by only a few days (or, if kept cold, a few weeks). Sterilization lengthens the storage life indefinitely but, at least in the older methods of treatment, produces a brownish colour and a caramelized flavour. To avoid this effect methods involving ultra-high temperature sterilization for a very short time have recently been introduced.

It is important to realize that a sterilized food can easily become recontaminated with microbes and go bad. If it is to keep well, it must be protected from recontamination in its own can or bottle, once the container has been opened the food can spoil.

Much food poisoning is caused by dirt, i.e. microbes from dirty hands, dirty dishes and dirty utensils, being put on the food after it has been sterilized by cooking or canning.

RE-HEATING

Cold slows down growth of microbes, sufficient heat kills them. Re-heated dishes should be properly heated right through (i.e. 170 °F, 77 °C) and not just kept warm—in the warmth microbes will grow at a fantastic rate. Keep food cold or very hot, but not warm.

Canning A sterilized food must be protected from recontamination with micro-organisms. Canning and bottling are processes by which the food is sterilized after sealing into its protective covering. The process was first introduced by Nicholas Appert nearly 50 years before Pasteur showed that fermentation and putrefaction are caused by living organisms which can be destroyed by heat.

In the canning process there is a balance between the minimum amount of heat needed to sterilize the food, and the maximum that can be applied without affecting the flavour and appearance of the food. For example, it is not possible to sterilize a 6 lb can of rice pudding because by the time that the middle of the mass of food has reached sterilizing temperature the outer part has become brown through over-heating. Food technologists have recently overcome this problem by sterilizing the food before canning (the canning must then be done under aseptic conditions).

There is some loss of vitamin C and a smaller loss of vitamin B_1

when foods are canned, but after that they retain their nutritive value for a long time. The vitamin C may fall gradually after a few months storage but the other vitamins and the protein are stable for years.

Drying The spoilage organisms cannot grow in the absence of moisture therefore another method of preserving foods is to dry them. Drying is one of the oldest methods of food preservation and has for centuries been done simply by exposing the food to the sun and the wind. The same principles are involved in modern drying methods but the conditions are closely controlled.

The old method of drying foods in the open air was very slow and usually resulted in poor appearance and flavour, and in the destruction of vitamin C and much of the B_1 (and often in bacterial contamination from dirt). In the factory the process is carried out in ovens or tunnels, or in streams of air, at controlled temperatures. The food is dried more quickly and retains more of the fresh flavour, colour and nutritive value. Further improvements have led to drying in vacuum, where lower temperatures are needed to evaporate the water and so less damage is done to the food.

Today a variety of equipment is used for drying different foods. The term 'dehydrated food', which means no more than 'dried food', tends to be used for foods dried under controlled conditions.

Solids such as meat, fish, vegetables and fruits are dried in hot air or in vacuum in ovens or on moving belts. Liquids may be dried on hot drums—'roller-drying'—or by spraying. In roller-drying the drum is heated from the inside, usually by steam, to a temperature above the boiling point of water. A thin film of the liquid is applied to the rotating drum and it evaporates within a few seconds. The drum rotates against a scraping knife so that the thin film of dried food is continually removed. As drying is completed within a few seconds, only a little damage is done from the point of view of attractiveness of the food and nutritional value.

In spray-drying the liquid is sprayed in fine droplets into a chamber of hot air, the temperature of which is well above the boiling point of water. Within a second or two the droplet has evaporated and a particle of dried powder falls to the bottom of the chamber. In such a short heating period only a little damage is done to the food.

Milk is dried by both of these methods and there is little to choose between the two products. Fruit juices, with their high sugar content, egg and 'instant' coffee and tea, give a better product when spray-dried.

Among the more recent of the drying methods is that called 'accelerated freeze-drying'. The food is frozen and then placed in a vacuum. The ice evaporates from the solid form without liquefying while the food remains frozen. The food has never been heated so that flavour, texture, colour and nutritional value are usually better than those of foods that have been dried by other methods. Actually gentle heat may be applied to the frozen food—this is the '*accelerated*' part of the process—but the rapid evaporation of the ice cools the food sufficiently to keep it frozen solid. Another virtue of the method is that the food is spongy in structure, retaining its shape, and so rehydrates within a few minutes. Foods dried by other methods often require several hours for rehydration.

Another method of drying is 'puffing'. The food, such as cereal grains or diced vegetables, is heated and then the pressure is suddenly reduced. The water inside the food evaporates very rapidly and the steam bursts its way out. The dried vegetable cubes are hollow and rehydrate fairly quickly. When the method is applied to cereal grains it results in breakfast cereals such as puffed wheat and puffed rice.

Smoking Wood smoke deposits chemicals on the surface of the food which act as preservatives. At the same time the surface dries out slightly. The result is that there is a protective coat around the food that helps to preserve it—but only for a limited time. Smoked foods such as kippers, bloaters and ham will keep for only a short time.

Jams Microbes cannot grow in the absence of moisture and for the same reason they cannot grow on a strong sugar syrup. The syrup holds the water in such a way that it is not available to the microbes. This is the principle of preserving fruit by making jam. It is essential that the sugar concentration is kept up to 70%; below this concentration yeast and moulds will grow on the fruit. An opened jar of jam can pick up moisture from the air so that the top layer is not sufficiently concentrated in sugar and the jam goes mouldy. This is why a circle of waxed paper is dropped on the surface of the jam after filling the jar.

The prolonged boiling involved by preserving fruit by this method destroys part of the vitamin C; there is a little left unless copper pans are used and then *all* the vitamin C is destroyed.

Pickling Just as micro-organisms cannot grow in the absence of moisture, many of them cannot grow in the presence of acid. This explains why pickling is a method of preserving. A salt pickle is far from sterile but is free from unwanted bacteria. The salt prevents the growth of harmful organisms and permits the growth of other organisms that are able to ferment the carbohydrate in the food into lactic acid. The acidity so developed both prevents the growth of spoilage organisms and confers the pickled flavour on the food, e.g. pickled cucumbers and sauerkraut. Vinegar may be used instead, e.g. pickled onions.

Chemical preservatives The spoilage of food by oxidation of the fat may be prevented by protecting it from the air or by adding chemicals that stop the process, i.e. antioxidants. An example of one of these chemicals that is legally permitted in many countries is butylated hydroxytoluene, or BHT. Many vegetable fats contain natural antioxidants which enable them to be stored much longer than animal fats, and one of these is vitamin E. This may therefore be added to fatty foods as a preservative.

Two chemicals that are generally permitted in fruit juices to prevent fermentation, are sulphur dioxide and benzoic acid (in strictly controlled amounts). Boric acid has been used in the past but is not permitted nowadays.

ENZYMES

Enzymes are naturally present in the tissue of all plants and animals. After the death of the animal or the removal of the vegetable material from the living plant, the enzymes continue to act and, in some cases, cause changes in the food. One example is the 'hanging' of game. This is traditional practice in Great Britain where grouse and pheasant are left to 'hang' at room temperature for several days, while their own enzymes soften the tissue and cause flavours to develop. In many foods, however, enzyme action causes undesirable changes in the food and must be stopped. (Such is the strength of tradition that if the 'hanging' process were applied to chicken it would be considered unpleasant.) Another example of enzyme action is the brown discolouration of a sliced apple.

Vegetables, which are usually a good source of vitamin C, contain an enzyme, ascorbic acid oxidase. When a food such as cabbage is harvested and allowed to wilt, or is chopped, the enzyme is able to come into contact with the vitamin and destroy it (see p. 45). Other enzymes can produce rancidity, protein degradation and the development of unpleasant flavours or the loss of the natural colour of the food.

Since they are proteins, enzymes can be destroyed by heat. They are not much affected by cold and so foods kept in the refrigerator may still deteriorate slowly although bacterial action has been greatly inhibited. For this reason foods such as fruits and vegetables which are to be stored by drying, and sometimes when they are to be stored by deep-freezing, are first immersed for about half a minute in boiling water, i.e. blanched, to destroy enzymes.

PROCESSING AIDS

A large number of substances are used in rather small amounts to assist in food processing. These are sometimes referred to as 'chemicals in foods' but as all the nutrients are chemicals this term is rather misleading and it is more correct to call them processing aids or chemical aids. Some of these are simple substances that every cook uses, such as salt, saltpetre, sugar and vinegar, others are more complex substances used to keep cakes moist (humectants), to slow down the staling of bread (anti-staling agents), or to help fats to spread or salad creams to remain stable (emulsifiers and stabilizers). In addition many colours and flavours are added to foods to make them more attractive or to restore the losses that often occur when they are preserved.

In addition there are other substances that are necessary if food is to be manufactured on a large enough scale to feed millions of people in our cities. For example many foods are now made by automatic machinery and the customer demands that every sample, day after day, should be of the same quality. As most of the raw materials are natural substances that vary from batch to batch it is often necessary to add processing aids to help to maintain this standard. We expect our daily loaf of bread to be the same although the flour will vary considerably from one delivery to the next. The baker may add enzymes to help him and also treat the flour with various chemicals to make it suitable for his process. 'High Ratio flour', treated with chlorine, can be used in cake-making with as much as 140 parts of sugar to 100 parts of flour, whereas ordinary flour can take up only half this quantity of sugar.

The addition of non-food substances to foods sometimes causes concern among members of the public as to their possible harmful effects. While there is no absolute safety there is certainly no cause for alarm as the use of these additives is strictly controlled by law in most countries. Only those considered safe by the medical authorities may be used and then only in controlled amounts. The reason why there is no *absolute* safety is that it is very difficult to prove that minute amounts of these substances taken daily over a lifetime do not have some effect—it is always difficult to prove a 'negative'. Although there is no absolute proof of harmlessness of everything added to food, at the same time there is no evidence that people in Great Britain have ever suffered any harm from processing aids.

WHAT IS IN THE PACKET?

People know what to expect when they buy vegetables or fruit or meat but with several hundred manufactured foods sold in packets it is not so obvious what we are buying. However, the Labelling Act ensures that the names of all the ingredients in the food are printed on the packet *in descending order of the quantities used.* This last fact is not generally known. Nor is it well known that we have a range of Food Laws designed to protect the consumer, and enforced by the Local Authorities. Consumer Protection Officers, Environmental Health Officers and Trading Standards Officers protect our food supplies and ensure that (a) they are not injurious to health, (b) they contain whatever nutrients the manufacturer claims, (c) no misleading claims are made, and (d) the Food Laws are obeyed.

In some instances the composition of the food is laid down by law, for example the amount of fruit juice in cordials, of fruit in jam, of meat and fish in breadspreads and pies, of water in butter, of fat in cream and dried milk.

Many people do not bother to read labels although there is a lot of information there. For example one may imagine that prepared tomato soup contains only tomato but the label may also list sugar, wheat flour, salt and fat, as food ingredients, and flavouring and colours such as sodium glutamate, spices, herbs, and natural or synthetic dyes.

Information of nutritional value can be gleaned from the label. Apart from specific vitamin claims, discussed below, a manufactured food may list, say, milk powder as an ingredient. If this is high up in the list it means that the product is more nutritious. A meat or chicken cube may list salt as the first and therefore the main, ingredient, followed by a dozen other ingredients. There are chicken curries with wheat flour as the first-named ingredient, and pie fillings with sugar as the main ingredient. If the housewife and caterer read the labels they will know what they are buying.

VITAMIN VALUES

To prevent manufacturers making claims for vitamins that are present in amounts too small to be of any nutritional importance, no claims may be made unless a 'reasonable' amount is present. This is not actually a legal enforcement but is a Code of Practice in Advertising and if a manufacturer disregarded it he could certainly be stopped under the regulation that prevents misleading advertising.

No claims may be made that a food contains a vitamin unless at least one-sixth of the daily allowance is present in an average portion. The food may not be claimed as a rich source unless an average portion contains at least half of the daily allowance.

For the purposes of this Code of Practice the daily allowances are laid down as being 3 000 i.u. vitamin A, 1 mg vitamin B_1, 1·8 mg vitamin B_2, 12 mg nicotinic acid, 30 mg vitamin C and 500 i.u. vitamin D; 0·75 g calcium, 0·1 mg iodine, 10 mg iron, 0·75 mg phosphate (1947).

The manufacturer takes care that the food contains the quantities of nutrients stated on the label, and to make sure of this he often provides a slight surplus. If the vitamin is likely to lose potency during storage he puts an expiry date on the packet. This does not mean that the food is in any way harmful after this date but only that there is no guarantee that the vitamins are still up to the label potency. Foods that lose flavour or go rancid on long storage may also carry expiry dates.

NUTRITIONAL LOSSES IN FOOD PROCESSING

It is popularly believed that while there is considerable destruction of nutrients during commercial manufacture, there is very little damage through home cooking. Not only is this far from true, but sometimes the reverse may be true. That is to say quality control can be exercised in food factories so that losses could be (but are not always) kept to a minimum, while some housewives are neither expert cooks nor nutritionists.

Many experiments have been carried out that indicate the type and extent of losses that *can* take place but there is little evidence available of the losses that, in practice, *do* take place. While it is possible to purchase and analyse fresh or stale or processed foods from shops and markets we do not know the nutritive value of meals eaten in the home. These may have been stored some time at home, cooked well or badly, and kept hot for varying periods before being eaten, or even reheated the next day.

The losses given as examples on Table 34 must be viewed against the following background:

1. Some losses are inevitable. Any wet process will leach out some part of the water-soluble nutrients and the amount lost depends on the size of the pieces of food (e.g. shredded or large pieces of leafy vegetables, small peas, large beans, etc.) and on the time of processing and the volume of water used. Frozen and dehydrated foods are always 'blanched' first to destroy any enzymes present and this is the stage where the greatest losses occur through leaching into the blanching water.

The trimming of vegetables and fish and meat, and the milling of wheat, all entail loss of nutrients.

TABLE 34
Examples of processing and cooking losses

Thiamin	*Loss*
Canned ham	40%
Roasting and stewing mutton	30%
Peas, boiling	15%
canning	60%
drying and boiling	70%
Potatoes, boiling	25%
steaming, baking, frying	10%
Riboflavin	
Peas, boiling	25%
canning	50%
drying and boiling	50%
Potatoes, boiling	25%
steaming, baking, and frying	no loss
Nicotinic acid	
Peas, boiling	40%
canning	50%
drying and boiling	60%
Potatoes, boiling	30%
steaming, baking, and frying	no loss
Vitamin C	
Fruit, stewing	30–40%
canning	30–40%
Potatoes, peeling and boiling	30–50%
boiling without peeling	20–40%
baking in skin	20–40%
frying	25–35%

During winter storage potatoes lose about one-sixth of their vitamin C every month.

2. There is some loss of almost all nutrients, but this is usually quite small, except for vitamin C, which is by far the most sensitive of the nutrients, and, to a lesser extent, thiamin.

Methods of reducing losses of vitamin C are discussed on page 48.

3. Cooking must be regarded as part of the processing; much of the loss that occurs in manufacture is *instead of*, not in addition to cooking losses. For example, canned meat or vegetables may need only heating before consumption, and canned fruit can be eaten immediately since the food is already cooked in the can. So the factory loss is instead of the kitchen loss. It may be greater or less than would have occurred in

the kitchen, depending on the skills both of the manufacturer and the housewife.

As a result very often there is little difference between a freshly cooked home dish and a processed one.

In one experiment fresh peas that needed 10 min. boiling, frozen peas that needed 3.5 min., and freeze-dried ones that needed only 2 min. finished up, respectively, with 16, 14, and 16 mg of vitamin C per 100 g. In the same experiment air-dried peas that needed 15 min. boiling finished up with 11 mg of vitamin C per 100 g and canned peas that had suffered losses in processing finished up with 9 mg.

4. It is important to consider if the food is a *significant* source of nutrients in the diet, e.g. partial destruction of vitamin C during pasteurization of milk is of limited importance since milk is not usually a significant source of this nutrient. Potatoes, however, are a major source of vitamin C and of thiamin in many diets so that damage here could be serious.

5. We often attempt to compare fresh and processed foods but 'fresh' meaning 'unprocessed' can be several days or weeks old, while the processed equivalent may have been treated within an hour or so of collection from the soil or the sea. So it is possible to find processed foods of *higher* nutrient content than the freshly cooked equivalent. We need to know whether the food is 'garden fresh' or merely 'market fresh'. The only nutrient that is lost to any extent during storage in the market is vitamin C.

It is important in this connection to realize that even very fresh foods, both animal and vegetable, vary considerably in nutrient content. Food composition tables merely provide representative figures.

6. Losses must be balanced against advantages. For example, when minced meat products, such as sausages and pies, are preserved with sulphur dioxide, thiamin is lost, but the consumer is protected from food-poisoning bacteria. On the other hand the loss of thiamin incurred when ready-made chipped potatoes are kept white with sulphur dioxide may *not* be considered by everyone to be a price worth paying for convenience.

7. It is not always a choice between processed food and fresh food but between processed food and none at all. This is simply because we could not have certain foods out of season unless they were preserved in some way.

8. Finally, there are beneficial effects of processing, including the arrest of any further loss of nutrients. For example, fruits and vegetables continually lose vitamin C during storage, even under the best condi-

tions, but if they are dried, frozen, or canned the nutrients are stable after the initial processing loss.

Heat processing destroys toxins in foods, such as soya beans and other legumes, cassava, and ackee (a West Indian fruit); treatment with alkaline lime water releases the niacin in maize, and makes it available to the body (see p. 19). Niacin is actually formed in coffee beans during roasting; vitamin C and E are added to some foods as processing aids, and so increase the nutritional value.

Lastly, the cook often destroys nutrients in producing an attractive flavour. For example, the flavour of baked, roasted, and fried products is largely due to the Maillard or browning reaction, in which part of the lysine of the proteins combines with reducing substances in the food. A visible example is the development of the brown colour as butter beans are being cooked. Flavour improves as the brown colour develops and the nutritive value of the protein slowly falls. It is not likely that this is of serious nutritional importance but, as mentioned earlier, we know very little about the nutritional value of food on the plate.

PASTEURIZATION OF MILK

The pasteurization of milk is an essential public health measure although there is some small loss of nutrients—clearly, a price worth paying for safe milk. There is no significant change in the protein quality nor in the vitamins A, B_2 or nicotinic acid. There is a loss of about 10% of the thiamin, and 20% of the ascorbic acid. Since milk is not a rich source of thiamin and ascorbic acid these losses are not important. Sterilization destroys 30% of the vitamin B_1 and 50% of the vitamin C.

NOTE: STABILITY OF VITAMINS

Vitamin A and carotene Destroyed by heat in the presence of air, especially in light and in the presence of copper, and to a lesser extent, iron. Stable to high temperatures in the absence of air; can withstand boiling in alkaline solution but not in acid. Destroyed by rancid fats. Being fat-soluble vitamin A and carotene are not extracted into the cooking water.

Thiamin Stable to acids, even at boiling point, but unstable under neutral or alkaline conditions. Therefore destroyed during baking if an alkaline baking powder is used. Unstable to air, destroyed by sulphur dioxide (often used as preservative); leached out into cooking water.

Riboflavin Stable to air and to acid but not to alkali. Destroyed by excessive heat and by light.

Nicotinic acid Most stable of the B vitamins, being unaffected by light, heat, air, acid or alkali. The only losses are by leaching into the water during processing or cooking.

Vitamin C Destroyed by enzyme in wilting leafy vegetables and in chopped vegetables and fruits. Unstable to air and the oxidation is accelerated by heat and copper. Small traces of copper can completely destroy the ascorbic acid in solution. Stable in canned and bottled foods. Readily leached into the cooking water. Protected by sulphur dioxide.

Vitamin D Stable to all normal storage and processing conditions.

ENRICHMENT

A variety of nutrients is added to foods to increase their nutritional value, either because such nutrients are legally enforced as a public health measure or because they are voluntarily added by food manufacturers. Most foods intended for babies and young children are enriched with a range of vitamins, iron, and calcium and sometimes with extra protein. Fruit juices are often enriched with vitamin C and breakfast cereals with thiamin, riboflavin, and niacin, as well as iron. Salt is enriched with iodine in certain areas (as iodide or iodate) and special 'slimming foods' often contain a complete range of nutrients.

PUBLIC HEALTH ENRICHMENT

So far as public health measures are concerned the procedure should follow a series of logical stages, although, historically this has not always been done.

The logical sequence of stages in enrichment would be: (1) demonstrate the need to enrich; (2) establish the best food to carry the nutrients; (3) ensure that there is no detriment to palatability and acceptability of the food, and that the cost is small; (4) develop the technology to carry out the enrichment; and (5) have the legal machinery to enforce the law. (This means inspectors to purchase samples, and analytical facilities on a large scale, as well as the legal machinery. It is this last difficulty that has sometimes prevented enrichment in developing countries most in need.)

Each of these stages needs to be examined in detail:

(1) The need for enrichment is obvious in developing countries, but in other communities where enrichment is common it is a precaution rather than a proven need. The enrichment of white bread in the United States followed reports of inadequate intakes of thiamin but there was no evidence of deficiency. The enrichment of flour with calcium in Great Britain in 1942 accompanied the higher extraction rate of flour in the expectation that the higher content of phytic acid would reduce calcium absorption. At the same time milk was in limited supply.

(2) The vehicle for enrichment must be one of the cheapest foods most commonly eaten by those at risk, i.e. the poorer sections of the community. So bread or rice is the commonest vehicle.

Enrichment cannot be carried out unless the nutrient is compatible with the food. For example a fat-soluble vitamin cannot be added to an aqueous food (although this particular problem has been overcome by the preparation of water-dispersible forms of the fat-soluble vitamins.)

(3) There must be no change in colour, flavour, texture, or general appearance, otherwise the enriched food will be rejected in favour of ordinary variety. Obviously the whole programme would fail in its intention if enrichment made the food much more expensive.

(4) It needs great technical skill and machinery to blend a few milligrams of a vitamin mix with many tonnes of flour or to enrich whole rice or wheat grains. Lack of technology often prevents enrichment in communities that need it most. It is not practicable to enrich foods that are processed in very large numbers of small factories, as is the case in many developing countries.

(5) There is little use in passing laws that cannot be enforced. Samples must be purchased by inspectors, analytical facilities spread throughout the country are essential, and there must be adequate legal machinery. Inability to enforce the law prevents such public health measures in many countries.

EXAMPLES OF ENRICHMENT

In Great Britain flour has been enriched since 1952 with the following (calcium enrichment was started in 1942): 0·24 mg thiamin, 1·6 mg niacin, 1·65 mg iron, and 145 mg calcium per 100 g white flour.

The British Food Standards Committee (1974) recommended discontinuing the addition of niacin since it is made in the body from tryptophan. The amounts of additives vary in different countries; Riboflavin is sometimes added.

Margarine is required by law in the U.K. to be enriched with vitamin

A to the highest levels found in summer butter, and with vitamin D namely 9000 μg retinol equivalent, and 70–90 μg vitamin D per kg (255 μg retinol equivalent per oz and 2·27 μg of vitamin D per oz) Many other countries also enforce enrichment of margarine.

The earliest enrichment, about 1900, was the addition of iodide to chocolate in Switzerland to prevent goitre, although a commoner vehicle for iodine is salt. Milk is enriched with vitamin D in the United States but not in Great Britain. Tea is enriched with vitamin A in parts of India in an attempt to prevent the widespread occurrence of xerophthalmia, and in Guatemala sugar is the vehicle for vitamin A.

CONTROL OF ENRICHMENT

Subject to provisions listed above, enrichment is relatively easy, but many countries do not allow it without Government permission and control of amounts added. Some countries ban all enrichment apart from the restoration of the milling losses from wheat.

Enrichment is sometimes useful as a short-term measure until eating habits can be improved by nutrition education. Enrichment of proprietary foods may be filling a need, or merely be a selling advantage. It is possible to distinguish between:

(a) Restoration of processing losses (e.g. in milling wheat to white flour).

(b) Enrichment, meaning addition of nutrients to a legally defined standard.

(c) Fortification, meaning the addition of nutrients which are not present in the food before processing or are present in only small amounts.

(d) Nutrification, an American term meaning the supplementation of snack foods so that they make a special contribution to the diet. This becomes a matter of importance where people tend to replace properly balanced meals with snack foods which may supply little more than energy.

In practice the terms 'enrichment' and 'fortification' are often interchanged. The addition of calcium to flour is fortification, since there is very little calcium in whole wheat; the addition of iron, thiamin, and niacin is restoration of milling losses, partly or completely; where riboflavin is added this is fortification as well as enrichment since there is little riboflavin in wheat. Since these additions are legally enforced the term enrichment would be the correct one to use.

FOOD LAWS

Food legislation is necessary:

(1) to prevent adulteration;
(2) to protect the consumer from harmful additions;
(3) to protect the consumer from bacteriological and toxic contamination;
(4) to prevent misleading claims;
(5) to establish and maintain compositional and nutritional standards.

The first general food laws anywhere in the world, as distinct from specific ones dealing with bread and beer, were passed in Great Britain in 1860. These were designed mainly to prevent adulteration and harm to the consumer. The principal intention of the law nowadays is to establish and maintain standards for the composition of foods, to provide information on the label, and to restrain unwarranted and misleading claims and advertisements. The law also controls the use of food additives.

Legislation in Great Britain is initiated by the Ministry of Agriculture, Fisheries and Food, sometimes in conjunction with the Department of Health and Social Security. It is advised by two independent committees, the Food Standards Committee, which advises on composition, labels, and advertising claims, and the Food Additives and Contaminants Committee, which lays down types and amounts of colours, flavours, texturizers, and other additives which may be used. This Committee, when necessary, works in conjunction with the Pharmacology sub-committee of the Committee on Medical Aspects of Food Policy of the Department of Health. This latter body would also be involved in food enrichment.

Enforcement of the law is in the hands of the local authorities via the local inspectorate and public analyst. So it can be seen that there are many safeguards of the quality of food and protection of the consumer, which are virtually continuously under revision and amendment. Similar laws protect the consumer in most countries although, of course, there are considerable national differences in the ways in which the law is operated. Consumers associations play a valuable part both in providing the public with information and in acting as a spur to legislators and enforcement officers.

Labels are an important source of information to the consumer and allow the housewife, so long as she understands them, to make a wiser choice of manufactured foods. Labels must list all the ingredients of the food in descending order of quantity (major ingredient first) together with a description of the contents and the name and address of the manufacturer in case of complaint.

FOOD STANDARDS AND DEFINITIONS (Great Britain)

BREAD

White bread—is defined as being made from flour, yeast and water and other additions including salt, fat, milk, sugar, enzyme preparations, rice flour and soya flour (not more than 2% of the flour), antimould agents, emulsifiers, bleaching or improving agents, wheat gluten and germ, poppy and caraway seeds, oatmeal (not more than 2% of the flour).

Brown bread or **wheatmeal bread**—must contain at least 0·6% fibre and may contain caramel and up to 5% soya or rice flour as well as the additions permitted in white bread.

Wheatgerm bread—must contain not less than 10% added processed wheatgerm.

Wholemeal bread—is made from wholemeal flour only, yeast and water and may contain other additions.

Milk bread—must contain not less than 6% milk solids and state 'skimmed' if such is the case.

Protein bread—must contain not less than 22% protein.

Gluten bread—must contain not less than 16% protein.

The term *starch-reduced* may not be used unless the food contains less than 50% carbohydrate on a dry weight basis.

Loaves that weigh more than 10 oz must be sold in multiples of 14 oz.

White flour must contain, per 100 g, not less than 1·65 mg iron, 0·24 mg thiamin, 1·6 mg nicotinic acid, and not less than 235 mg and not more than 390 mg of chalk. Self-raising flour must contain not less than 0·25% calcium.

MILK

Pasteurized milk has been treated to a temperature of not less than 161 °F for 15 seconds. *Sterilized milk* has been homogenized and heated in the bottle to a temperature of 212 °F. *UHT* (Ultra-heat treatment) *milk* has been subjected to a temperature of 270 °F for 1 second.

Channel Islands and *South Devon milk* contain at least 4% butter fat. *Single cream* or *coffee cream* contains 18% fat; *sterilized cream*, 23%; *double cream* and *clotted cream*, not less than 48% fat. *Butter* and *margarine* may not contain more than 16% water.

MEAT PRODUCTS

These are governed by various regulations.

Meat with jelly, luncheon meat and *meat with cereal* should contain not less than 80% meat. *Meat loaf* or *meat roll* should contain not less than 65% meat. *Pork sausages* should contain 65% meat, and beef sausages 50% meat.

Meat in this context means the flesh of the animal together with the skin, gristle, sinews and fat naturally associated with it.

FRUIT DRINKS

Fruit crush is defined as a soft drink ready for consumption without dilution, whereas a *squash* is a drink that needs dilution. *Cordial* can be either.

A *crush* must contain at least 5% juice by volume, at least 4½ lb of added sugar per 10 gallons and may contain not more than 56 grains of saccharin per 10 gallons. Lime crush need have only 3% citrus juice.

A *squash* must contain at least 25% fruit juice, at least 22½ lb of sugar per 10 gallons and may contain not more than 280 grains of saccharin per 10 gallons. Non-citrus squashes need have only 10% fruit juice.

A drink made from whole fruit, including flesh, pith and rind, i.e. a comminuted product, must contain at least 2 lb of fruit per 10 gallons if it is to be drunk without dilution and at least 10 lb of fruit if it is to be diluted. Sugar and sweetening as for crush and squash.

'Diabetic' soft drinks must not contain any added sugar and may contain any quantity of artificial sweetener.

Low-calorie soft drinks must not contain more than 7½ kcal per fl oz if prepared for drinking after dilution and not more than 1½ kcal if ready to drink.

The amount of preservative must not exceed 350 parts per million of sulphur dioxide or 800 p.p.m. of benzoic acid. If the preparation is to be drunk without dilution these quantities are reduced to one-fifth.

JAM AND PRESERVES

Jam must contain not less than 68½% soluble solids and not less than a fixed percentage of fruit, depending on the type of fruit, e.g. rhubarb—40%, blackcurrants—25%, others—30 or 40%. (These figures vary because of the different setting power of the fruits.)

Marmalade must contain not less than 20% citrus fruit; *fruit curd*, not less than 4% fat, 0·33% citric acid, 0·125% oil of lemon or 0·25% oil of orange, 1½% dried egg or 1% dried egg white (or 4½% shell egg).

FOOD ADDITIVES

Only designated foods may contain preservatives. These include bread, cheese, raw fish, fruit, jam, pickled meat, sausages, soft drinks and beer, etc. The permitted preservatives include sulphur dioxide, benzoic acid, propionic acid, sorbic acid, sodium nitrate, orthophenyl phenol, etc.

Mineral oils such as liquid paraffin are not permitted in any foods except dried fruit, sugar confectionery, cheese, eggs and chewing compounds in limited quantities.

The use of certain antioxidants for fats is permitted within prescribed limits. These include gallates of propyl, octyl, and dodecyl alcohol, BHT (butylated hydroxytoluene) and BHA (butylated hydroxyanisole).

The use of emulsifiers and stabilizers includes glycerol esters, sorbitan esters of fatty acids, stearyl tartrate, cellulose derivatives, and brominated vegetable oils.

EXAMINATION QUESTIONS

1. As a catering manager would you advocate the use of canned, frozen, dried and dehydrated fruits and vegetables? Give detailed reasons for your answer. (R.S.H. Certificate.)
2. How does deterioration of foods come about, and what causes it? Briefly outline the methods used in the home to preserve fruit and vegetable, emphasizing in each case, the special care required to ensure that preservation will be successful. (City & Guilds 244.)
3. What part do the prepared packet and frozen foods play in a modern household? Discuss the advantages and disadvantages of some of the foods. What care is required in buying, storing and using them? (City & Guilds 244.)
4. Comment on the nutritive value in relation to cost, of potatoes, dried carrots, frozen peas, baked beans in tomato sauce, and canned blackcurrants. (R.S.H.)
5. What are the ingredients and nutritive qualities of (a) ice-cream, (b) soups, (c) breakfast cereals?
6. What are the differences in nutritive value between low-calorie drinks for slimmers and those for diabetics, and starch-reduced and high-protein breads?

7. With the aim of reducing catering costs it is suggested to you as a caterer, to use the following:
   (*a*) Margarine instead of butter; (*b*) fish fingers instead of fresh fish; (*c*) dried peas instead of frozen peas; (*d*) canned fruits instead of freshly stewed fruit; (*e*) frozen beefburgers instead of home-made ones. Discuss these suggestions giving the factors you would consider in making your decision. (R.S.H.)
8. You have been asked to assess the following products for home use for convenience and nutritive value: (*a*) mashed potato powder; (*b*) dried skimmed milk; (*c*) dried yeast; (*d*) frozen puff pastry; (*e*) frozen concentrated orange juice; (*f*) packet mushroom soup. Prepare six dishes, each dish using one of the products. Choose recipes with special emphasis on good food value together with quick and easy preparation. Calculate the energy value of one dish. (Cambridge 'A' level practical Home Economics.)
9. The contents of a packet mix are stated as follows: skimmed milk powder, edible starch, salt, wheat flour, edible fat, monosodium glutamate, sugar, dehydrated onion, hydrolysed protein, dehydrated mushrooms, yeast extract, emulsifier, flavouring, preservatives, antioxidant. Write notes on four of the ingredients underlined with special reference to their functions, chemical nature, and properties. What type of product would you expect when the appropriate liquid is added. Give reasons. (Welsh Joint Board 'A' Level 1975.)
10. Describe the processing involved in four of the following and discuss the advantages to the consumer of being able to purchase them:
    (a) refined sugar, (b) ultra-heat-treated milk, (c) accelerated freeze-dried peas, (d) quick-frozen chicken, (e) refrigerated vacuum-packed sliced ham, (f) dehydrated potato-mash powder. ('A' Level 1973.)
11. To what extent and in what ways can advertising be said to affect the choice of diet in this country? (London 'A' Level 1976.)

9 World food problems

There has never been enough food available to feed the whole human family properly. It is only in recent years that the entire population even of the industrialized countries has had enough to eat. Historians tell us that in the first 1800 years of the present era there was a famine one year in every five. Even in Great Britain, where food is nowadays so plentiful, there was a recorded famine one year in every ten.

Mechanized farming, the liberal use of fertilizers, irrigation and pest control, together with the application of food technology in preserving food between harvests have, in recent years, abolished food shortages in the industrialized countries. The shortages remain, however, in the non-industrialized countries.

FOOD & AGRICULTURE ORGANIZATION

At the end of the Second World War the Food & Agriculture Organization of the United Nations was set up to examine, report on and advise on these problems. Since then FAO has given, and continues to give, technical advice on all aspects of food production—fishing, farming, animal husbandry, food manufacture—to all countries that request it. In some of its activities FAO acts in partnership with the United Nations Children's Fund, which has the special responsibility of the welfare and nutrition of children, and with the World Health Organization.

Figures of food production and population statistics are not available for every country, nor are they very reliable, but FAO has concluded that a tenth of the world's population, that is 300 to 500 million people go hungry. That is to say they do not get enough calories to satisfy their needs: they are undernourished. In addition there are about a thousand million people who are not getting the right kinds of food: they suffer from malnutrition. In many areas the main shortage is protein, in others one or more of the vitamins may also be in short supply.

DIETARY SHORTAGES

It is good nutritional advice to eat a little of everything. In this way our

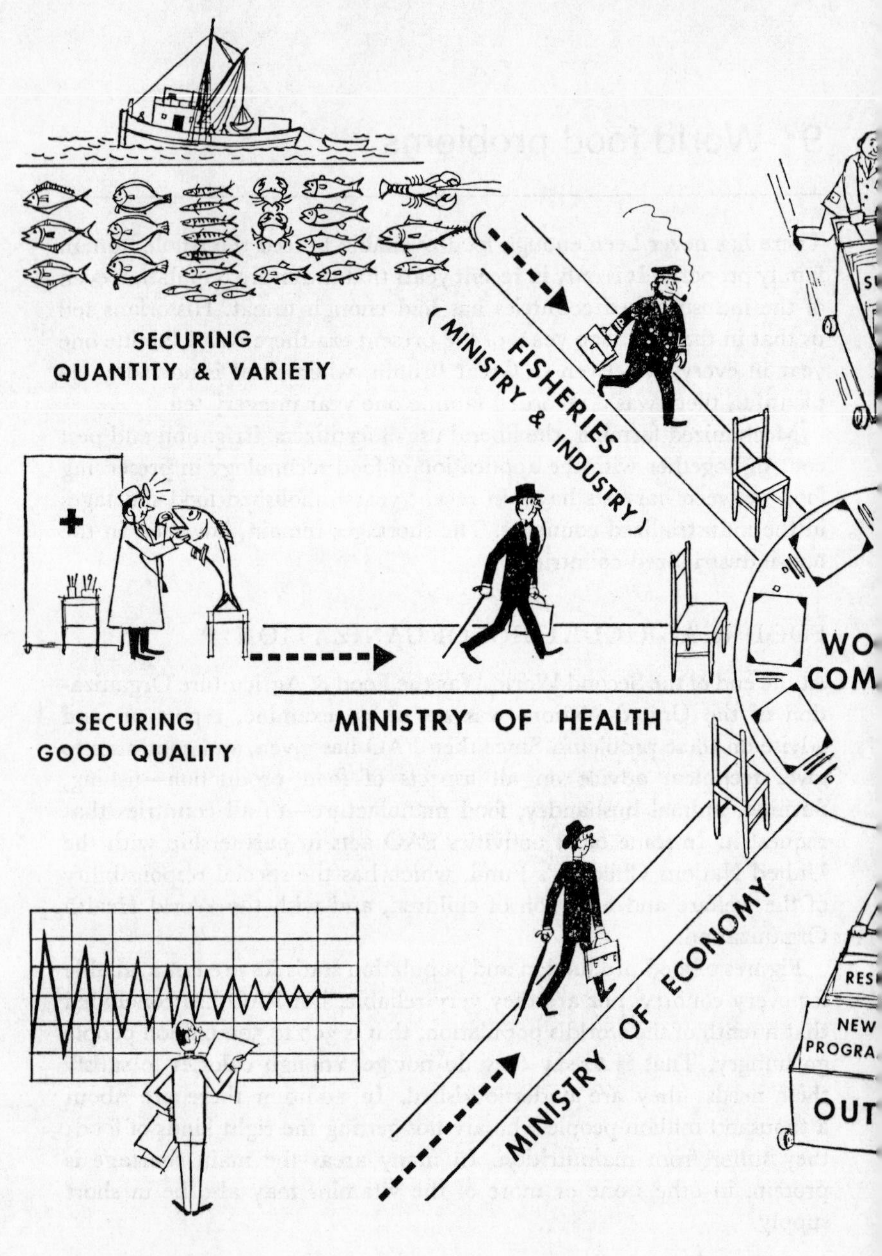
PROMOTING AVAILABILITY
SECURING QUANTITY & VARIETY
FISHERIES
(MINISTRY & INDUSTRY)
SECURING GOOD QUALITY
MINISTRY OF HEALTH
STABILISING PRICES
MINISTRY OF ECONOMY
WO
COM
RES
NEW
PROGRA
OUT
THE COMPOSITION OF A COMMI

Fig. 24 Promoting availa

CREATING A DEMAND

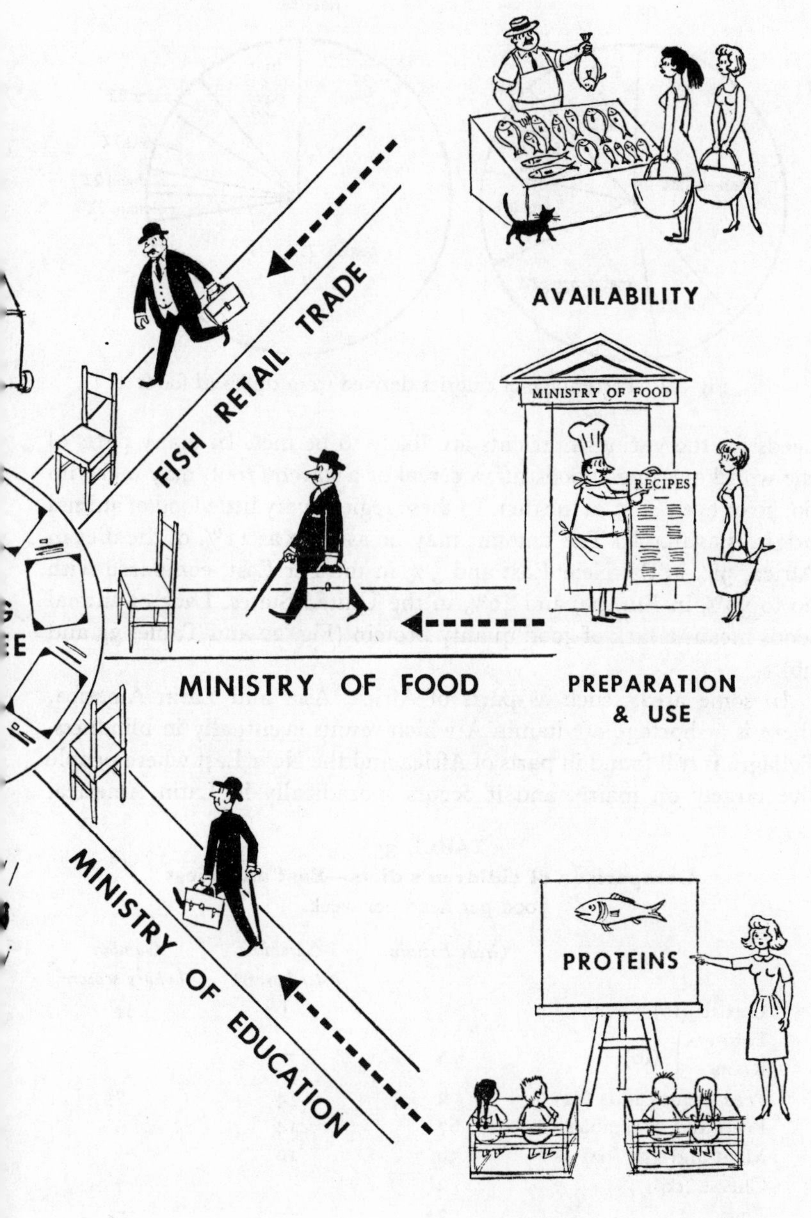

E IN RELATION TO ITS WORK

creating a demand

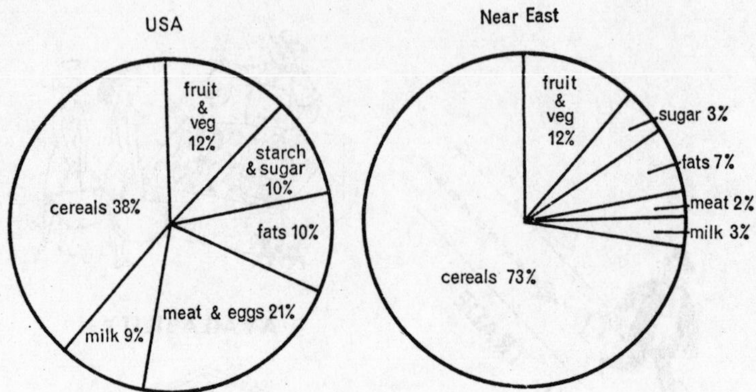

Fig. 22 Percentage of calories derived from different foods

needs for the various nutrients are likely to be met. In many parts of the world one single foodstuff, a cereal or a starchy root, may make up 60, 70 or even 80% of the diet. In these regions very little food of animal origin is available. The amount may be as little as 11% of the diet in Africa, 9% in the Near East and 5% in the Far East, compared with 20 to 30% in Europe, and 40% in the United States. Lack of animal foods means a lack of good quality protein (Fig. 22 and Tables 35 and 36).

In some areas, such as parts of Africa, Asia and Latin America, there is a shortage of vitamin A which results eventually in blindness. Pellagra is still found in parts of Africa and the Near East where people live largely on maize, and it occurs sporadically in Latin America.

TABLE 35
Comparison of children's diets—East and West
Food per head per week

	Great Britain	*Gambia after harvest*	*Gambia hungry season*
Cereals (lb)	5	6¾	5½
Potatoes, Roots (lb)	5	—	—
Peas, beans, nuts (oz)	2	14	7½
Fruit and vegetables (oz)	67	14	5
Meat and fish (oz)	36	10	—
Cheese (oz)	2½	—	—
Eggs	2½	—	—
Milk (pints)	5½	—	—
Fats (oz)	12½	¾	—
Sugar (oz)	4	—	—

TABLE 36
Diets of 5-year-old children

United Kingdom		*S. and E. India*	
Breakfast		Rice	6 oz (170 g)
Cornflakes	½ oz (14 g)	Brown sugar	2 oz (57 g)
Milk	5 oz (140 g)	Dried fish	½ oz (14 g)
Bread	1 oz (28 g)	Vegetables	
Butter	½ oz (14 g)	*Total*	
Bacon	1 oz (28 g)	kilocalories	**900**
Tomato	2 oz (57 g)	protein	**25 g**
Midmorning orange juice	5 oz (140 g)		
Dinner		*Ghana, village near Kumasi*	
Steak	2 oz (57 g)	Plantain	2½ oz (70 g)
Carrots	2 oz (57 g)	Cocoyam	8 oz (230 g)
Potatoes	2 oz (57 g)	Cassava	1½ oz (42 g)
Milk pudding	5 oz (140 g)	Beans and nuts	⅕ oz (5 g)
Tea		Cocoyam leaf	½ oz (14 g)
Egg	2 oz (57 g)	Fish and meat	⅓ oz (10 g)
Bread	2 oz (57 g)	Red palm oil	
Butter	¾ oz (20 g)	Egg plant	
Jam	½ oz (14 g)	Red peppers	
Apple	4 oz (114 g)	Onions	
Milk	5 oz (140 ml)		
Bedtime			
Milk	5 oz (140 ml)		
Total			
kilocalories	**1 500**		**430**
protein	**60 g**		**11 g**

Beri-beri is common among people eating polished rice. Nutritional anaemia due to iron deficiency is the chief cause of maternal mortality in developing countries.

PROTEIN-ENERGY MALNUTRITION

Protein-energy malnutrition (PEM) describes a range of disorders, from marasmus at one end, which is a severe and continuous shortage of food, to kwashiorkor at the other, where the energy supply may be nearly adequate but protein is deficient. Between the two extremes are various disorders based on a deficiency of energy and protein as well as minerals and vitamins, together with associated infections.

PEM is responsible for deaths of large numbers of young children in the developing countries, where death rates are 20 to 50 times higher than in the West and where as many as half the children born die before reaching 5 years of age.

Marasmus means to waste (from the Greek) and is partial starvation. It occurs mostly in children under 12 months of age living in towns. The babies are very small, wizened, and shrunken with no subcutaneous fat.

Kwashiorkor is the name given by the Ga tribe of Ghana to the disorder that occurs during the second year of life, mainly in rural areas. The body is small, may have some subscutaneous fat, but is swollen with oedema. Usually an infection precipitates the disorder.

Both disorders follow early weaning from the breast on to unsuitable foods. Alternatively a baby may be breast-fed without any supplements for 1½ to 2 years and so gradually becomes more and more undernourished.

PEM is not necessarily associated with poverty, but often with ignorance. Food and the money to buy it may be available, but the mother has very little knowledge about infant feeding. Custom and taboo may deprive the child of meat, or fish, or eggs, or even of milk, when these are available. Poor hygiene, particularly in bottle-fed infants, leads to frequent bouts of gastroenteritis which aggravate the condition. Such children are very susceptible to infections, such as measles and whooping cough, which can prove fatal to them, whereas properly fed children almost always recover.

In areas where the staple foods—for example cassava, yams, plantains, and sweet potato—are low in protein adults can manage with the addition of small amounts of peas and beans to supply additional protein. The babies suffer for two reasons. First, such protein-rich foods as are available may be witheld from children because it is believed that eggs or meat or similar foods are 'not good' for them. Secondly, since a baby has a proportionately higher need of protein, diets suitable for an adult may be inadequate for them. Another cause is staple foods that are too bulky to supply enough energy and protein. For example, maize appears from analysis to contain enough protein (8–10%) but maize meal swells up with water to quite a large volume and the baby's stomach is simply too small to hold enough food for its needs.

PROTEIN-RICH PREPARATIONS

These children need food not medicine. Much work has been devoted, largely by the United Nations Children's Fund, to preparing protein-rich mixtures suitable as weaning foods. In many areas there are protein foods available which are not usually used for food, for example fish meal and oil-seed meal. Fish meal is dried fish and contains up to 70% protein of high biological value (if properly dried). The flavour can be removed by washing with various solvents and the white tasteless powder, called fish flour, may be added to bread for older children and adults, or incorporated into porridges and paps for babies.

There are several seeds which are produced mainly for their oil content such as groundnut, linseed, sesame and cottonseed but which are also rich in protein. After removal of the oil the residue, the oil-seed cake, contains about 40 to 50% protein of moderate to good biological value.

One of the first of the infant protein-mixtures made from vegetable sources was Incaparina. This was named after the initial letters of the Institute of Nutrition of Central America and Panama. It consists of maize and the cereal sorghum, together with cottonseed flour, which is rich in protein, with addition of a small amount of food yeast as a source of B vitamins, and calcium carbonate and vitamin A. The cereals are cooked, ground, and mixed with the other ingredients. The mother mixes the food with 10 parts of water and cooks it for 15 minutes.

In several countries there are similar mixtures based on oilseeds, fish flour or milk powder with cereals, minerals and vitamins.

In India a large amount of groundnut meal is available and this is used as the basis of several protein mixtures. One of these is called Indian Multipurpose Food (IMF) and consists of 75% groundnut flour and 25% Bengal gram. Another preparation is called Nutro macaroni because it looks like macaroni but is made from wheat semolina and groundnut flour with added calcium, iron and vitamins.

These protein-rich foods are designed for babies because the greatest need is for weaning foods and used for all young children who have high requirements for protein.

At the same time as the preparation of these protein-rich mixtures is being encouraged, the Food & Agriculture Organization is fostering the production of protein-rich foods such as peas and beans, and of animal protein foods by improving egg, meat and milk production and by exploiting fishing grounds.

MORE PEOPLE—MORE FOOD

The problem of food supplies is intensified by the rapidly increasing population. As medical care saves lives, the child mortality falls and the total population grows. In the bad old days so many children died that the population remained constant—death-rate equalled birth-rate.

Then as the death-rate was reduced and while the birth-rate continued at its same high level, the size of the population increased rapidly—the 'population explosion'. This is shown in the graph (Fig. 23)—in 1830 the world population was one thousand million. It took 100 years to add another thousand million—two thousand million by 1930. It took only 30 years to add the third thousand million. In the next 30 to 40 years, by the year 2000, yet another three thousand million people will have been added to bring the total up to six to seven thousand million.

The increases are not evenly distributed in all countries but are greater in the developing countries. This is because the high death-rate is steadily being reduced there while the birth-rate stays high; whereas in the industrialized countries the death-rate has already been improved and birth control restricts the numbers. For example, the death-rate of children in the first year of life in some of the developing countries is

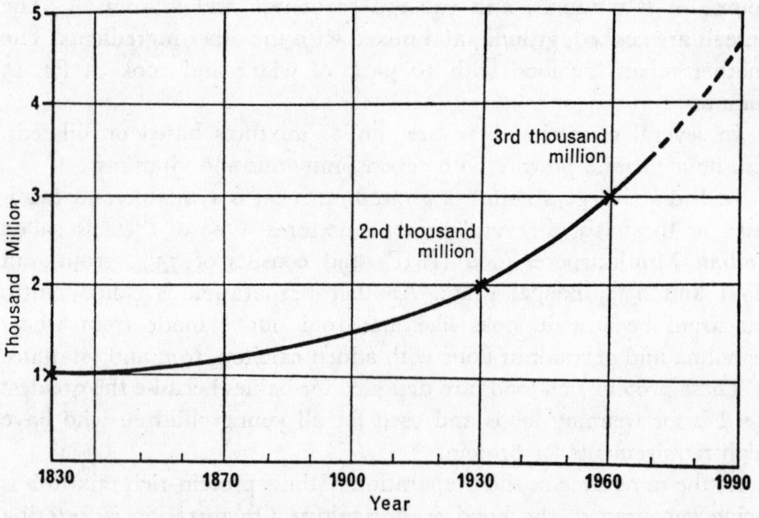

Fig. 23 Rapid increase in world population

often ten times that of the West, and in children up to the age of five it can be forty times as much. Obviously, as these figures are improved, and they are steadily improving, the population in those countries will expand enormously.

Those countries where improvements are taking place are ones that are already short of food. The shortages must become more severe unless there is a tremendous increase in food production and unless, at the same time, the best nutritional use is made of everything available and the population increase is slowed down (see Fig. 23).

NOVEL FOODS

This term is applied to foods that have either not been previously eaten at all or have been eaten only in small amounts. Since they are mostly rich in protein and could make a contribution towards solving world food problems they can be dealt with here, although at present they are mostly being used by caterers and food manufacturers in developed countries.

Novel foods may, purely for convenience, be classed as primary foods, substitute foods, micro-organisms, foods from non-food sources, and synthetic foods.

(1) **Primary foods** are those eaten at first hand, as distinct from those eaten as products from animals that ate the primary food in the first place.

There is nothing novel about vegetarianism, but novelty arises from the direct consumption by man of primary foods like plankton, on which fishes feed, and krill, on which whales feed. These are available in gigantic quantities compared with the limited amount of seafood that is available. For example, it is calculated that it takes 10 000 kg of plankton to produce 1 kg of fish. Among the many problems to be overcome before such foods are available is the high energy cost of collecting them.

Another primary food is leaf protein. Green leaves contain 1·5–5% protein and are plentiful in the wild or can be repeatedly cropped as grass. Their indigestible cellulose, however, severely limits the amount that we can eat, unless the protein is extracted. The process is simple. The leafy material is crushed and the juice squeezed out and the protein coagulated by heat. It is dark green in colour and has a strong 'grassy' flavour, liked by some people and disliked by others. It has been tried out in some areas by mixing it with other foods or special dishes and has, so far, had a mixed reception.

(2) **Substitute foods** The most significant of these is textured vegetable protein (TVP) in which mainly soya, (although other proteins can be used) is texturized and flavoured to resemble meat. This is being used as a meat replacer or extender for cheapness. Certainly, in terms of eating vegetable food directly instead of feeding it to animals, soya products make more efficient use of the land. Soya can be made to resemble other foods like coconut and raisins, but the main use is as a meat-like product and many appetizing and savoury dishes can be prepared from it.

In case TVP does eventually replace a major part of the meat in our diet, legislation is being discussed to enforce the addition of nutrients to make it nutritionally equivalent to meat. Rehydrated TVP has about the same amount of protein as meat but with slightly lower quality—net protein utilization about 50–60% instead of 75%, but this is not important in the diet as a whole. It has less thiamin than meat and no vitamin B_{12} and it is suggested that these be added. Furthermore, the iron in meat, haem iron, is particularly well absorbed compared with other dietary sources and it is suggested that this be compensated for by adding extra iron.

Substitute foods could be improved beyond the nutritional value of the food they replace. Apart from adding nutrients in excess of those in meat, the fat used in preparing the product could be richer in polyunsaturates than meat fat, if this should be considered desirable, or dietary fibre or mineral salts could be added.

Coffee whitener (or 'creamer') is another example of a substitute or imitation food. In this instance ease of handling and long storage life are the advantages over milk, although it is greatly inferior in nutritional value.

(3) **Micro-organisms** Algae, bacteria, moulds, and yeasts have all been grown in tanks on a factory scale. They offer the advantage of controlled production at a very rapid rate. For example scaled down to one day:

a steer weighing 1000 kg would produce 1 kg of protein;
1000 kg of soya beans would produce 86 kg of protein;
1000 kg of yeast would produce 4000 kg protein; and.
1000 kg bacteria would produce 10 million kg protein.

All this without use of soil and without dependence on the weather.

Yeast is already in large-scale production in factories producing tens of thousands of tonnes a year for animal feed using petroleum by-products or carbohydrate waste-products as a source of energy. Examples of the latter are waste from paper-making and waste from pineapple canning. Small amounts are added to human foods both for flavour, which is rather strong, and as a source of B vitamins.

Bacteria are being produced on a large scale for animal feed from methanol prepared from methane from natural gas.

Algae, being plants, have the special advantage that they do not require a supply of energy since they obtain this from sunshine and need only mineral salts, a source of nitrogen, and carbon dioxide. A microscopic form, Chlorella, has been investigated in great detail but not developed for food.

A blue–green alga called Spirulina, long eaten by the Aztecs and in parts of North Africa, is being investigated because it obtains its nitrogen from the air instead of requiring ammonium or other nitrogen salts to be added to the growth medium.

Seaweeds are also members of the alga family and are eaten in some parts of the world (e.g. laver bread in South Wales) but it is the unicellular form that offers possibilities of large-scale factory production. In fact yeasts, bacteria, and algae (and also incorrectly, moulds) are usually called single cell protein or SCP.

The mould Fusarium can be grown on waste starch products and is being developed for human food, whereas the other main SCP developments, namely yeast and bacteria, are intended for animal feed.

(4) **Food from non-food sources** Examples of these are glucose and alcohol from wood waste from timber mills and paper manufacture, and protein from waste wool and hydrolysed hog bristle.

These are not being developed but serve as examples of the possibilities of obtaining food supplies from waste materials.

(5) **Synthetic foods** Fats and carbohydrates have been synthesized, but the processes cannot be compared with crop cultivation. Vitamins nowadays are mostly produced synthetically as are many of the flavours, colours, and texturizers added to foods. The amino acids lysine and methionine are being synthesized on the factory scale for supplementation of the proteins of animal feeds.

These novel foods illustrate possible future developments that may become necessary to feed the growing population. In general they cannot (yet) compete in price or attractiveness with conventional foods conventionally grown. Greater success has been achieved by the geneticists who have produced new varieties of plants and animals with higher yields.

The problem is that most people prefer familiar foods. They do not want to eat strange foods just because they may be surplus from another country (like dried skimmed milk), or cheaper, or more convenient to produce.

So problems of food shortages are made good by providing more of of the familiar foods, or money to buy them, or by improving agriculture, for example the Green Revolution.

The Green Revolution refers to the recent successful breeding of new varieties of rice, maize, and wheat which respond particularly well to irrigation and fertilizer and produce crops two and four times as great as were grown before.

Care must be taken to ensure that colour, flavour, and texture are fully acceptable; this means they must be like the foods usually eaten.

EXAMINATION QUESTIONS

1. What types of malnutrition are found in the developing countries and what are the reasons?
2. Why are babies particularly vulnerable to food shortages? What are the possible remedies?
3. Why is there so much emphasis on protein-rich preparations for infants in the developing countries?
4. It is often said that a great part of the problem of world food shortage could be solved by making better use of foods that are available. Comment on this.
5. What sources of protein could be made more available in the developing countries? Make some suggestions for their practical use which would appeal to the people.
6. Discuss the balance between population control and agricultural improvement in relieving world food shortage.
7. A particular country is attempting to promote increased fish consumption, to alleviate protein shortage in the national diet. Suggest ways of popularizing fish.
8. Say what you know about FAO, UNICEF, WHO, and their works.
9. Which international agencies are concerned with problems of food shortage and how do they help?
10. What nutritional deficiencies are likely to occur in countries where people live largely on cereals? What practical steps can be taken to improve the nutritive value of the diet?
11. Why is dried skimmed milk so greatly esteemed as a dietary supplement? Suggest ways of promoting its use in tropical countries.

GENERAL QUESTIONS

1. Do you think that the slogan 'Drinka Pinta Milka Day' is a good one? If so, why? Some children will not drink milk as such. Give a week's menu for a child aged five years, showing what you would do under such circumstances. (R.S.H.)
2. Describe the ways in which the government intervenes to increase the nutritional value of the British diet. (R.S.H.)
3. You are the Catering Manager of an industrial canteen serving lunches daily for 600 manual workers and 80 administrative staff in separate dining-rooms, with all cooking being carried out in one kitchen.

(a) Give an example of the type of menu you would offer in each dining-room.

(b) State the size of the cooked portions you would serve and estimate the quantities of raw ingredients required for each menu.

(c) Calculate the cost of ingredients at present-day prices, find the cost per head and state the suggested selling price.

(d) Suggest what type of service you would operate, giving reasons for your choice. (Hotel & Catering Institute, Final Membership.)

4. Discuss the advantages and disadvantages of the use of Textured vegetable protein in welfare catering. (R.S.H.)

5. Discuss the use of food tables in assessing the nutritive value of diets. (R.S.H.)

6. 'Cheap foods are always inferior in quality.' Discuss this statement in the light of your catering and nutritional knowledge. (R.S.H.)

7. A hospital caterer discovers that his costs are too high. He is asked to reduce them. Suggest ways of doing this, without decreasing nutritional values. (R.S.H.)

8. An office girl eating in a cafeteria chose the following lunch:

   Baked beans (2 oz), bread as toast ($1\frac{1}{2}$ oz), butter ($\frac{1}{4}$ oz), bun (3 oz), coffee, with 2 oz milk.

   Plan suitable meals for this girl for breakfast and supper and calculate the protein, calories and vitamin C available from all three meals. (R.S.H.)

9. Give the three main principles that should always be followed in planning meals to ensure that they are balanced nutritionally. What size portion of the following foods would you serve to an adult woman, and what would they each contribute to her diet?

   (a) Frozen peas (c) Wholemeal bread
   (b) Bacon (d) Stewing steak

   (Hotel & Catering Institute.)

10. 'The ideal diet for man would consist of plain, uncooked foods, eaten in their natural state, without sophistication by the modern food manufacturer.' Write (a) arguments in favour of this contention, (b) arguments against it, and (c) your own verdict. Separate (a) from (b) and (c) in your answers. (R.S.H.)

11. Recent nutritional research has put forward the view that 'the average business man eats far more than he needs. By today's standards the right amount of food needed by people in sedentary jobs is so low as to be almost ridiculous.' Discuss this from the point of view of the Catering Manager in Hotels and Restaurants.

What sort of foods would be appropriate for meals on trains used largely by business men? Illustrate your answer with some typical menus.

12. Large scale institutional catering produces nutritional problems. Which of these do you consider to be most important and how would you try to overcome them?
13. How would the nutritional requirements of a middle aged company director and an athletic young Hotel and Catering student vary? Suggest a suitable day's menu for each and indicate the approximate cost. (Hotel & Catering Institute, Intermediate.)
14. Give an account of the factors which contribute to good nutrition and physical well-being. What is the importance in the diet of (a) water and (b) roughage? (City & Guilds Adv. Dom. Cookery.)
15. What is the importance of raw foods in the diet? Give menus for a midday meal in (a) summer, (b) winter, in which raw foods are included. (City & Guilds 243.)
16. 'The health of the nation is in the hands of the housewife.' Discuss this statement. (City & Guilds 244.)
17. A Health pamphlet recommends adults to eat this group of foods daily and then eat what they like:
    $\frac{1}{2}$ pt milk, six slices of bread (5 oz), one egg (2 oz), one orange (4 oz) or $\frac{3}{4}$ pt of juice, 1 oz cheese, $1\frac{1}{2}$ oz oatmeal or lean bacon, $\frac{1}{4}$ oz cocoa or 1 oz plain chocolate, plus a weekly portion of liver (4 oz) and either $\frac{1}{2}$ lb margarine or, 100 g tin of sardines.
    Calculate:
    1. What proportion of the recommended intakes of calcium and vitamin C this group of foods provides.
    2. The cost of the foods each week.

       Prepare and serve, using foods from the group
       (*a*) a slimming lunch for three women;
       (*b*) a substantial breakfast for a very active man;
       (*c*) two sweet and two savoury dishes;
       (*d*) an appetising dish using liver. ('A' level Practical Home Economics.)
18. How would you decide if a food is a good source of iron? How should a Home Economist promote a convenience food which is a poor source of this nutrient? Illustrate your answer with three recipes for printing on prepacked ground rice or cornflour. (Isleworth Polytechnic Home Economics.)
19. Recent surveys show an increasing number of children going to school with an inadequate breakfast or none. Why is this a matter

for concern? Say what you consider are the essential components of breakfast for school children, justifying your statements. Proceeding from your suggestions plan the rest of the day's intake showing how the full dietary requirements can be met. (London 'A' Level 1973.)

20. Discuss the need for and the nutritional significance of the following practices: (a) including dietary fibre in the diet; (b) using polyunsaturated fats; (c) eating yoghurt; (d) taking synthetic vitamin supplements. (London 'A' Level 1977.)

Conclusion

Good health and physique depend on many factors, such as heredity, hygiene, exercise, climate, but the most important factor is the food we eat.

Nutrition is concerned with food, its production, processing, distribution, preparation, nutrient requirements, and their final utilization in the body. The end-product of practical nutrition is a good choice of food, on the plate, in the shopping basket and in the store cupboard. It is a quantitative subject, like fuelling a car. It is true that recommended allowances of nutrients and tables of food composition refer to *average* groups of people and foods, but they are as useful as average prices as a guide for feeding human beings, and help us to get enough of the foods which are the 'best buys' for nutritive value.

THE CORE FOODS PLAN

The nutrition 'evidence' for good food choice is a lengthy scientific study, but the 'verdict', namely the actual foods chosen, can be simple. For every country nutritionists can select a small group of foods in quantities which cover most of the nutrients needed to ensure a satisfactory diet. These core foods can be chosen from the lists of food portions previously given. The best foods to choose will be those which give good food value in an eatable portion, are popular, available and reasonably cheap. The total planned food value should provide about two-thirds of the recommended allowances for nutrients, but only a small amount of energy, which can always be safely left to individual choice. The foods chosen in each group will vary with local tastes and availability but they will have the same nutritive value: they can be described as *rations for health*.

In a tropical country—where the staple foods are white rice, yams, cassava, which supply chiefly energy, this group of core foods, eaten regularly, would ensure a satisfactory diet whatever else was eaten.

Figures are from *The Composition of Foods in Tropical Countries*.

25 g dried skimmed milk	50 g pulses
15 g dried fish or 25 g river snails	50 g peanuts
50 g egg	25 g red palm oil

Vitamin C-rich fruits, such as paw-paw and guavas are eaten daily and no vitamin D would, of course, be needed. This food group would provide:

	Nutrient totals	*Fraction teenage allowance*	*Fraction adult allowance*
Protein	50 g	$\frac{2}{3}$	$\frac{4}{5}$
Calcium	800 mg	all	150%
Iron	6–7 mg	$\frac{1}{2}$	$\frac{2}{3}$
Vitamin A	1 500 μg	all	all
Thiamin	0·9 mg	$\frac{3}{4}$	$\frac{3}{4}$
Riboflavin	0·8 mg	$\frac{1}{2}$	$\frac{1}{2}$
Nicotinic acid	16 mg	$\frac{7}{8}$	all

In Great Britain similar nutrient totals would be supplied by a group of different foods:

Daily

1 oz cheese (25 g)
$\frac{1}{2}$ pint milk (300 ml)
1$\frac{1}{2}$ oz bacon or oatmeal (40 g)
7 oz (200 g) mixed white and brown breads
4 oz orange, fruit (AP) or juice canned
10 g cocoa or meat extract

Plus weekly

1 portion of liver or kidney, and fat fish e.g. sardines (3$\frac{1}{2}$ oz (100 g) E.P.) and 7 oz (200 g) margarine.

	Nutrient totals	*Fraction adult allowance*
Protein	47 g	$\frac{2}{3}$
Calcium	780 mg	150%
Iron	6–7 mg	$\frac{2}{3}$
Vitamin A	3 000 μg	3 times
Thiamin	0·8 mg	$\frac{2}{3}$
Riboflavin	1·5 mg	all
Nicotinic acid equivalent	20 mg	all
Vitamin C	50 mg	all
Vitamin D	3 μg	satisfactory

For teenagers and pregnant mothers, more milk, cheese and breads would be necessary, e.g. 1 pint milk, 2 oz cheese and 10 oz breads, with *two* portions of offals and rich fish weekly. There is no such thing as a 'perfect' diet, but there are hundreds of good diets. For example, these two suggested diets are quite different, but supply the required nutrients. Cheaper foods could be chosen such as skimmed milk (dried or condensed) instead of fresh, brewers' yeast instead of bacon, raw cabbage instead of oranges; but it is better policy to choose and promote nutritious foods which people like and enjoy.

A TEACHING AID

The core food plan is a practical way of putting over better nutrition. For teaching purposes the foods can be put on a tray, or into a shopping basket as a visual aid to memory. Housewives and caterers can use it as a guide to shopping, choice of recipes and budgeting. The cost of the core foods is an index of the cost of a good diet.

At national level *rations for health* would be a useful method of promoting better nutrition, and could be used in school meals, maternity clinics and as a basis for welfare foods. Expert knowledge is needed to make the correct choice of foods in each particular instance, but the core food group method is easy to use and leaves people with a good deal of free choice in the rest of the diet.

The result of using *rations for health* would be to ensure better nutrition and well-being in the family and the nation.

Appendix

WEIGHTS AND MEASURES

BRITISH

16 ounces (oz)	= 1 pound (lb)
14 lb	= 1 stone
112 lb	= 1 hundredweight (cwt)
20 cwt (2 240 lb)	= 1 ton (long ton)
2 000 lb	= 1 short ton
1 000 kg	= 1 metric ton (tonne)
1 oz of water	= 1 fluid ounce (fl oz)
1 imperial (i.e. standard) pint	= 20 fl oz
1 reputed pint	= $13\frac{1}{3}$ fl oz
1 reputed quart (wine bottle, 1/6 gallon)	= $26\frac{2}{3}$ fl oz

METRIC EQUIVALENTS

1 oz	= 28·4 grams (g)	100 g	= $3\frac{1}{2}$ oz (approx.)
1 lb	= 0·45 kilograms (kg)	1 kg	= 2 lb 3 oz
1 pint	= 0·57 litres (l)	1 l	= $1\frac{3}{4}$ pt
1 yard	= 0·91 metres (m)	1 m	= 1·09 yd
1 inch	= 2·54 centimetres (cm)	1 cm	= 0·39 in
1 gallon	= 4·55 litres (l)	1 l	= 0·22 gal

1 ml (millilitre or cubic centimetre)
= 1/28 fl oz

TEMPERATURE

To convert °F to °C, subtract 32 ° and multiply by 5/9.
To convert °C to °F, multiply by 9/5 and add 32 °.

0 °F = −18 °C	200 °F = 93 °C
32 °F = 0 °C	212 °F = 100 °C
100 °F = 38 °C	250 °F = 121 °C

DEFINITIONS

Kilocalorie (kcal). The amount of heat required to raise the temperature of 1 kg water by 1 °C (more precisely from 14°C to 15°C).

1 kcal = 4 180 joules = 4·18 kilojoules (kJ) = 0·004 megajoules (MJ).

International units. Formerly used as measure of vitamins A and D.

1 microgram (μg) retinol = 3·33 i.u. vitamin A

1 μg cholecalciferol = 40 i.u. vitamin D

RECOMMENDED INTAKE OF NUTRIENTS

Individuals vary considerably in the amounts of the various nutrients that they require so it is not normally possible to state the requirements of a given subject. Instead, figures have been calculated for each nutrient such that if those amounts are consumed most individuals who are in good health will have their requirements satisfied. Such figures are variously called 'optimum requirements', 'recommended allowances' and 'recommended intakes'. The Food and Agriculture Organization of the United Nations prefers this last term. So far as is known there is no advantage in consuming quantities greater than these.

It is important to recognize that these figures are set high enough to take care of the variation between individuals—they are amounts which should satisfy the needs of the greater part of the population. So if a person is found to be consuming less than this he is not necessarily suffering from any deficiency as he may simply have smaller needs.

We have not sufficient knowledge to calculate these figures with any great accuracy, which accounts for the fact that they are frequently revised, and partly accounts for the variation between tables produced by different authorities. In addition some countries may have special reasons for suggesting different levels.

In this book, for the purpose of compiling menus, we have used F.A.O. figures where they have been established. To obtain a figure for the adult those for males and females have been averaged.

NICOTINIC ACID EQUIVALENTS

Nicotinic acid equivalents include the contribution made from the amino acid tryptophan; the content of the food is then free nicotinic acid plus $\frac{1}{60}$ of the tryptophan. Since much of the nicotinic acid of cereals is probably unavailable, U.K. authorities suggest that the nicotinic acid naturally present in cereals should not be included.

Dietary goals for the United States
(U.S. Senate Select Committee on Nutrition and Human Needs. Washington, D.C. 1977)

1. Increase carbohydrate as complex carbohydrates to 55–60% of energy intake.

2. Reduce total fat consumption from 40% to 30%.
3. Reduce saturated fats to 10% of energy intake and replace the fat with 10% polyunsaturated fats and 10% mono-unsaturated.
4. Reduce cholesterol intake to 300 mg per day.
5. Reduce sugar to 15% of the energy intake.
6. Reduce salt intake to about 3 g per day.

VITAMIN NOMENCLATURE

Vitamin	*Chemical names*
(General term)	(Specific substances)
A	Retinol, carotene
B_1	Thiamin
B_2	Riboflavin
Niacin	Nicotinic acid, nicotinamide (equally active)
B_6	Pyridoxine (or pyridoxol), pyridoxal, pyridoxamine
B_{12}	Cyanocobalamin
Folic Acid	Pteroyl-glutamic acid
C	Ascorbic acid
D	Cholecalciferol (natural form) Ergocalciferol (synthetic form)
E	4 forms of tocopherol; 4 forms of tocotrienol
K	Phylloquinone, menaquinones
Menaquinone	
H	Biotin
None	Pantothenic acid

Conversion table

used for changing oz to g (rounded-off totals)

1 oz = 28 g
2 oz = 57 g
3 oz = 85 g
3½ oz = 100 g
4 oz = 114 g
5 oz = 140 g
6 oz = 170 g
7 oz = 200 g
8 oz = 230 g
9 oz = 260 g

RECOMMENDED INTAKES—FOOD & AGRICULTURE ORGANIZATION

	Age years	*MJ*	*kcal*	*Protein* (g/kg)*	*Calcium (g)*	*Iron (mg)*	*Vitamin A (μg)*	*Thiamin (mg)*	*Riboflavin (mg)*	*Nicotinic acid (mg)*	*Vitamin D (μg)*
Children	0– 1	0·47	112/kg	1–3	0·5–0·6	7	300	0·4	0·6	6·6	10
	1– 3	5·7	1 360	1·19	0·4–0·5	7	240	0·5	0·7	8·6	10
	4– 6	7·6	1 830	1·01	0·4–0·5	8	300	0·7	0·9	11·2	10
	7– 9	9·2	2 190	0·88	0·4–0·5	10	390	0·8	1·2	13·9	2·5
Males	10–12	10·9	2 600	0·81	0·6–0·7	12	570	1·0	1·4	16·5	2·5
	13–15	12·1	2 900	0·72	0·6–0·7	15	720	1·2	1·7	20·4	2·5
	16–19	13·0	3 070	0·60	0·5–0·6	15	750	1·4	2·0	23·8	2·5
	20–30	12·6	3 000	0·57	0·4–0·5	10	750	1·3	1·8	21·1	2·5
Females	10–12	9·8	2 350	0·76	0·6–0·7	18	720	1·0	1·4	17·2	2·5
	13–15	10·5	2 490	0·63	0·6–0·7	15	720	1·0	1·4	17·2	2·5
	16–19	9·6	2 310	0·55	0·5–0·6	15	750	1·0	1·3	15·8	2·5
	20–30	9·2	2 200	0·52	0·4–0·5	20	750	0·9	1·3	15·2	2·5
pregnancy		+1·5	+350	+9	1·0–1·2	15	750	0·4/1 000 kcal	0·55/1 000 kcal	6·6/1 000 kcal	10
lactation		+2·3	+550	+17g	1·0–1·2	15	1200	ditto	ditto	ditto	10

* Chemical score 100.

UNITED STATES FOOD AND NUTRITION BOARD, RECOMMENDED DAILY DIETARY ALLOWANCES, REVISED 1973

	age from up to	weight		height		energy	protein	fat-soluble vitamins				water-soluble vitamins							minerals					
								vitamin A activity		vitamin D	vitamin E activity	ascorbic acid	folacin	niacin	riboflavin	thiamin	vitamin B_6	vitamin B_{12}	calcium	phosphorus	iodine	iron	magnesium	zinc
	yr	kg	lb	cm	in	kcal	gm	R.E.	← I.U.		→	mg	µg	←	mg		→	µg	mg	mg	µg	←	mg	→
Infants	0·0–0·5	6	14	60	24	kg.×117	kg.×2·2	420	1 400	400	4	35	50	5	0·4	0·3	0·3	0·3	360	240	35	10	60	3
	0·5–1·0	9	20	71	28	kg.×108	kg.×2·0	400	2 000	400	5	35	50	8	0·6	0·5	0·4	0·3	540	400	45	15	70	5
Children	1–3	13	28	86	34	1 300	23	400	2 000	400	7	40	100	9	0·8	0·7	0·6	1·0	800	800	60	15	150	10
	4–6	20	44	110	44	1 800	30	500	2 500	400	9	40	200	12	1·1	0·9	0·9	1·5	800	800	80	10	200	10
	7–10	30	66	135	54	2 400	36	700	3 300	400	10	40	300	16	1·2	1·2	1·2	2·0	800	800	110	10	250	10
Males	11–14	44	97	158	63	2 800	44	1 000	5 000	400	12	45	400	18	1·5	1·4	1·6	3·0	1 200	1 200	130	18	350	15
	15–18	61	134	172	69	3 000	54	1 000	5 000	400	15	45	400	20	1·8	1·5	1·8	3·0	1 200	1 200	150	18	400	15
	19–22	67	147	172	69	3 000	52	1 000	5 000	400	15	45	400	20	1·8	1·5	2·0	3·0	800	800	140	10	350	15
	23–50	70	154	172	69	2 700	56	1 000	5 000		15	45	400	18	1·6	1·4	2·0	3·0	800	800	130	10	350	15
	51+	70	154	172	69	2 400	56	1 000	5 000		15	45	400	16	1·5	1·2	2·0	3·0	800	800	110	10	350	15
Females	11–14	44	97	155	62	2 400	44	800	4 000	400	10	45	400	16	1·3	1·2	1·6	3·0	1 200	1 200	115	18	300	15
	15–18	54	119	162	65	2 100	48	800	4 000	400	11	45	400	14	1·4	1·1	2·0	3·0	1 200	1 200	115	18	300	15
	19–22	58	128	162	65	2 100	46	800	4 000	400	12	45	400	14	1·4	1·1	2·0	3·0	800	800	100	18	300	15
	23–50	58	128	162	65	2 000	46	800	4 000		12	45	400	13	1·2	1·0	2·0	3·0	800	800	100	18	300	15
	51+	58	128	162	65	1 800	46	800	4 000		12	45	400	12	1·1	1·0	2·0	3·0	800	800	80	10	300	15
Pregnant						+300	+30	1 000	5 000	400	15	60	800	+2	+0·3	+0·3	2·5	4·0	1 200	1 200	125	18	450	20
Lactating						+500	+20	1 200	6 000	400	15	60	600	+4	+0·5	+0·3	2·5	4·0	1 200	1 200	150	18	450	25

RECOMMENDED DAILY INTAKES OF ENERGY AND NUTRIENTS FOR THE UK (1969)

Age range	Occupational category	Body weight	Energy		Protein[2]	Thiamin	Riboflavin	Nicotinic acid	Ascorbic acid	Vitamin A	Vitamin D	Calcium	Iron
		kg	kcal	MJ[1]	g	mg	mg	mg equivalents	mg	µg retinol equivalents[3]	µg cholecalciferol	mg	mg
BOYS AND GIRLS													
0 up to 1 year		7·3	800	3·3	20	0·3	0·4	5	15	450	10	600	6
1 up to 2 years		11·4	1200	5·0	30	0·5	0·6	7	20	300	10	500	7
2 up to 3 years		13·5	1400	5·9	35	0·6	0·7	8	20	300	10	500	7
3 up to 5 years		16·5	1600	6·7	40	0·6	0·8	9	20	300	10	500	8
5 up to 7 years		20·5	1800	7·5	45	0·7	0·9	10	20	300	2·5	500	8
7 up to 9 years		25·1	2100	8·8	53	0·8	1·0	11	20	400	2·5	500	10
BOYS													
9 up to 12 years		31·9	2500	10·5	63	1·0	1·2	14	25	575	2·5	700	13
12 up to 15 years		45·5	2800	11·7	70	1·1	1·4	16	25	725	2·5	700	14
15 up to 18 years		61·0	3000	12·6	75	1·2	1·7	19	30	750	2·5	600	15
GIRLS													
9 up to 12 years		33·0	2300	9·6	58	0·9	1·2	13	25	575	2·5	700	13
12 up to 15 years		48·6	2300	9·6	58	0·9	1·4	16	25	725	2·5	700	14
15 up to 18 years		56·1	2300	9·6	58	0·9	1·4	16	30	750	2·5	600	15
MEN													
18 up to 35 years	Sedentary	65	2700	11·3	68	1·1	1·7	18	30	750	2·5	500	10
	Moderately active		3000	12·6	75	1·2	1·7	18	30	750	2·5	500	10
	Very active		3600	15·1	90	1·4	1·7	18	30	750	2·5	500	10
35 up to 65 years	Sedentary	65	2600	10·9	65	1·0	1·7	18	30	750	2·5	500	10
	Moderately active		2900	12·1	73	1·2	1·7	18	30	750	2·5	500	10
	Very active		3600	15·1	90	1·4	1·7	18	30	750	2·5	500	10
65 up to 75 years	Assuming a sedentary life	63	2350	9·8	59	0·9	1·7	18	30	750	2·5	500	10
75 and over		53	2100	8·8	53	0·8	1·7	18	30	750	2·5	500	10
WOMEN													
18 up to 55 years	Most occupations	55	2200	9·2	55	0·9	1·3	15	30	750	2·5	500	12
	Very active		2500	10·5	63	1·0	1·3	15	30	750	2·5	500	12
55 up to 75 years	Assuming a sedentary life	53	2050	8·6	51	0·8	1·3	15	30	750	2·5	500	10
75 and over		53	1900	8·0	48	0·7	1·3	15	30	750	2·5	500	10
Pregnancy, 2nd and 3rd rimester			2400	10·0	60	1 0	1·6	18	60	750	10	1200	15
Lactation			2700	11·3	68	1·1	1·8	21	60	1200	10	1200	15

[1] Megajoules (10^6 joules). Calculated from the relation 1 kilocalorie = 4·186 kilojoules, and rounded to 1 decimal place.
[2] Recommended intakes calculated as providing 10 per cent of energy.
[3] 1 retinol equivalent = 1 µg retinol or 6 µg β-carotene or 12 µg other biologically active carotenoids.

SPECIMEN CALCULATIONS OF THE NUTRIENT VALUES OF FOODS

Notes on the use of tables of food composition

Analyses are usually given for the edible portions of foods (E.P.), i.e. after removal of the non-edible part. Some analyses are given for food 'as purchased' (A.P.) e.g. the weight of a chop including the bone. Care must be taken to select the right analysis from the tables.

Example 1

Calculate the protein supplied by 7 oz (200 g) lamb chop, grilled, weighed with bone.

The Composition of Foods (McCance & Widdowson) lists five analyses of lamb chops; No. 276 is the one to use here: composition per 100 g—277 kcal, 18·3 g protein.

Example 2

Calculate the amount of protein and energy provided by 6 oz chicken, raw, weighed with bone.

In this example the tables given in the *Manual of Nutrition* (H.M.S.O.) are used. These show waste—31%, hence 6 oz chicken as purchased (A.P.) contains 69% of 6 oz = 4·2 oz edible portion; 1 oz E.P. supplies 41 kcal and 5·9 g protein, therefore 4·2 oz E.P. supply 170 kcal, and 24·8 g protein.

Note: Multiplication of 5·9 by 4·2 gives 24·78, but it would imply false accuracy to quote such figures to two decimal places. Similarly calorie totals should be calculated to the nearest 5, e.g., 4·2 × 41 = 172·2—this is entered as 170 kcal.

Example 3

Calculate the cost of 10 g of protein from lamb chops at 80 p per lb.

Use analysis number 274 (*Composition of Foods*, H.M.S.O. 1978) for lamb chops. The chops are 82% edible and contain 14·6 g protein per 100 g i.e. 4·1 g/oz E.P. Of the 16 oz purchased, only 82% is edible i.e. 13 oz (370 g).

Total protein in 13 oz chops is 13 × 4·1 i.e. 53·3 g protein, cost 80p.

Therefore 10 g of protein will cost $\frac{10 \times 80}{53 \cdot 3} = 15\text{p}$.

Example 4

Calculate the grammes of protein per penny provided by lamb chops at 80p per lb.

Use analysis number 274 as before. Lamb chops are 82% edible, contain 4·1 g protein/oz E.P.

80p buys $\frac{16 \text{ oz} \times 82\% \times 4{\cdot}1 \text{ g}}{100 \times 80\text{p}}$ = 0·66 g per penny or *0·7 g protein per penny.*

Compare this with the amount of protein per penny from minced beef at 54p per lb.

Use analysis number 247. There is no waste; protein content is 18·8 g per 100 g or 5·4 g per oz.

54p buys 16 oz × 5·4 g protein, so *one penny buys 1·6 g protein*

To find the 'Best Buy' for nutrients or energy, calculate and compare the amounts supplied per penny. This calculation is specially worthwhile for fruits and vegetables since they vary so much in vitamin C value—and also in waste.

Example 5

Which is the 'Best Buy' for vitamin C: green peppers at 30p per lb, or tomatoes at 21p per lb, or cabbage at 7p per lb?

Green Peppers are 84% edible, contain 28 mg vitamin C per oz E.P.

30p buys $\frac{16 \text{ oz} \times 84 \times 28}{100 \times 30\text{p}}$ mg i.e. 13 mg vitamin C per penny.

Tomatoes are 100% edible, contain 7 mg vitamin C per oz E.P.

21p buys $\frac{16 \text{ oz} \times 7}{21\text{p}}$ i.e. 5 mg vitamin C per penny.

Cabbage is 70% edible, contains 15 mg vitamins C per oz E.P.

7p buys $\frac{16 \text{ oz} \times 70 \times 15}{100 \times 7\text{p}}$ mg i.e. 24 mg vitamin C per penny, for raw cabbage.

Boiled cabbage, quickly cooked and served, will supply 9–10 mg vitamin C per penny.

(Manual of Nutrition.)

Example 6

Calculate the portion of a food supplying a given proportion of the recommended intake (R.D.I.) of a nutrient.

(a) How much milk supples half the R.D.I. of calcium for teenagers?
The R.D.I. of calcium for teenagers is 700 mg daily.
34 mg calcium supplied by 1 oz milk.
350 mg calcium will be supplied by 350/34 = 10 oz or ½ pint milk.

(b) Calculate the portion of baked beans supplying ⅓ of R.D.I. of iron for women.
The R.D.I. of iron for women is 12 mg daily.
0·4 mg of iron supplied by 1 oz baked beans.
4 mg of iron will be supplied by 4 mg/0·4=10 oz baked beans.

(c) Calculate a 100 kcal (0·42 MJ) portion of chicken on the bone.
Chicken E.P. supplies 41 kcal (171 kJ). Chicken is 70% edible.
Chicken A.P. supplies 70% of 41 kcal, viz 28·7 kcal (120 kJ).
100 kcal are supplied by 100/287=3·5 oz chicken on the bone.
420 kJ are supplied by 420/120 oz=3·5 oz chicken on the bone.
(120 kJ).
100 kcalories are supplied by 100/287 = 3·5 oz chicken on the bone.
420 kJ are supplied by 420/120 oz = 3·5 oz chicken on the bone.

N.B. *Finding the 'Best Buy' for nutrient value*; another way is to compare the costs of the various portions of foods each supplying, for example, ⅓ of R.D.I. of a nutrient.

FOOD PORTIONS SUPPLYING 10 GRAMMES OF PROTEIN

	oz	*grammes*
Dairy produce		
Cheese, Cheddar	1½	42
Cottage cheese	2	57
Eggs, whole	3½	100
yolk	2	57
white	4	114
Milk, fresh	½ pt	280 ml
" skimmed, dried	1	28
" " condensed	3½	100
Meat		
Bacon, Ham	2	57
Beef, stewing	2	57
Chicken, meat only	1½	42
Chicken with bone	3	85
Corned beef	1½	42
Luncheon meat	1½	42
Heart, kidney, liver	2	57
Lamb chop, with bone	3	85

Lamb, scrag end bone	4½	230
Minced beef	2	57
Sausages, beef, pork	4	114
Steak, grilling	2	57
Tripe	3	85
Fish		
Cockles, mussels	3	85
Coley/cod fillet	2	57
Cod, dried salt	½	14
Herring, whole	3½	100
Kipper, whole	3½	100
Herring, Kipper fillet	2	57
Pilchards, canned	1½	42
Salmon, canned	1¾	50
Roes, cod, soft	1¾	50
Shrimps, Prawns	2	57
Sprats whole	5	140
Almonds	1¾	50
Soya flour	1	28
Haricot beans, lentils	1½	42

FOOD PORTIONS SUPPLYING 100 Kcal (0·42 MJ)

*Foods marked with *** and ** contain zero or little carbohydrate*

10 g Oils***, lard***, suet***, dripping***

15 g Butter***, margarine***, mayonnaise, peanut butter**, shelled nuts**

20 g Crisps, chocolate, flaky and rich short pastry, sweet biscuits, cream crackers, shortbread

25 g Plain and water biscuits, cheese**, double cream**, lamb chop*** (+ bone), bacon***, ham***, oatmeal, rich fruit cake, sugar, salad cream, Victoria sandwich

30 g Crispbreads, scones, lunch meat**, dried skimmed milk, condensed milk, Edam cheese**, boiled sweets, breakfast cereals, flour, rice, pasta

35 g Cooked meats***, dry pulses, raw herring and kipper fillets***, sardines in oil***, jam, honey

40 g Bread, dried fruit, corned beef***, skimmed condensed milk

45 g Tinned pilchards***, salmon***, single cream**, soured cream**, chips

50–55 g Raw meats***, ice cream, spirits (50 ml)

60–65 g Fried fish**, eggs*** (whole, boiled or poached), evaporated milk

70–100 g Sherry (75 ml), cooked pulses (90 g), most soups (95 g)

100 g and over Cooked chicken on the bone***(100 g), baked beans (110 g), cottage cheese** (115 g), jelly** (120 g), boiled potatoes (125 g), shellfish*** (125 g), raw white fish fillet*** (140 g), table wines** (145 ml), fresh milk (150 ml), yoghurt (150 ml), grapes (165 g), raw tripe*** (180 g), cooked fresh peas (200 g), whole bananas (220 g), stout (260 ml), orange juice** (260 ml), whole apples (280 g), raw whole plaice*** (300 g), low fat yoghurt (300 ml), ale (330 ml), whole oranges** (370 g), cooked carrots** (500 g)

Practically NO energy Green vegetables, cucumbers, mushrooms, tomatoes, consomme, black coffee

Table 37

Desirable weights for men, aged 25 and over

Height (*without shoes*) ft	in	*Small frame*	*Medium frame*	*Large frame*
5	2	115–123	121–133	129–144
5	4	121–129	127–139	135–152
5	6	128–137	134–147	142–161
5	8	136–145	142–156	151–170
5	10	144–154	150–165	159–179
6	0	152–162	158–175	168–189
6	2	160–171	167–185	178–199
Desirable weights for women, aged 25 and over				
4	8	92–98	96–107	104–119
4	10	96–104	101–113	109–125
5	0	102–110	107–119	115–131
5	2	108–116	113–126	121–138
5	4	114–123	120–135	129–146
5	6	122–131	128–143	137–154
5	8	130–140	136–151	145–163
5	10	138–148	144–159	153–173

Food Preparation

When the value of foods is known, then the next stage is to collect and invent *recipes* which make good use of foods. Traditional recipes may be sound, or need modifying. New foods may need new types of recipes. Food preparation skills are a great help in making nutritious foods appetising and less good foods into nutritious dishes. Recipes should be evaluated for the preservation of nutritive values, consumers' tastes, cost and the equipment and fuel available.

RECIPES FOR SOME DISHES REFERRED TO IN THE TEXT

These three recipes for using small amounts of meat and fish, with rice, and local vegetables and flavours, come from China, Africa and Spain.

1. STIR–FRY MEAT. Takes 15 minutes to prepare and 5 minutes to cook. With a sharp knife, cut paper-thin slices of lean meat—pork, beef or any kind of liver. Mix the meat with 2 to 3 tablespoons of sherry or saké to each ½ lb of meat, 1 to 2 tablespoons of soy sauce, a teaspoon of cornflour, ½ teaspoon of ground ginger, ½ clove of garlic crushed with salt. This mixture is fried in 2 to 3 tablespoons of hot oil in a frying pan, over strong heat, stirring all the time with chopsticks, for 3 mins. Serve in a hot dish, garnished with salted nuts, chopped green spring onions, thin slices of lemon and peppers, with boiled rice.

   In another version of the dish, a red or green pepper, thinly sliced, is fried with the meat.

   This quick, dry method of cooking ensures tender protein, good protein value from the combination of cereal and meat, and practically *no loss* of B-vitamins and vitamin C.

2. JOLLOF RICE is an African dish. 'Jollof' means 'specially nice'. A meat stew is made in a pot by frying onions and tomatoes in oil with some meat, cut small, then adding salt, pepper and water to cover. Simmer for 1½ hours. Washed rice is cooked with water over low heat until all the water is absorbed and the rice is soft. One cup of rice absorbs 2 cups of water and serves two. The stew is mixed with the rice, some shellfish added such as mussels, oysters, and the mixture served in a hot dish garnished with

chopped green spring onion, peppers, shrimps, and sliced tomatoes.

3. PAELLA is the same mixture of meat, fish, rice and vegetables, but coloured with saffron, and cooked in one large iron pot. Fry the chopped garlic, onion, peppers, with meat cut small, in hot oil, then the rice and water are added with a pinch of saffron, and the mixture simmered for ½ hour. Some shellfish and pieces of fish, fresh (or frozen) peas and beans added and cooked with the rice for another ¼ hour. More water is added as needed. The rice should be moist but not 'soupy'. Add chopped parsley and serve.

YOGHURT is a thick fermented clotted milk, safe and digestible, good for infants and adults. It is made from fresh milk (boiled and cooled) or sterilized or UHT milk. For each 570 ml (1 pint) allow 150 ml (¼pt.) evaporated milk, or 50 g (2 oz.) dried milk, whole or skimmed, and a tablespoon of natural yoghurt (bought or home-made). The mixture is whisked or liquidized and poured into small pots, which have been well washed and rinsed with hot water to get rid of traces of detergent. Stand the pots in hot water in a casserole or saucepan with lid, leave 4–5 hours in a warm place, or in the hot sun. If not quite set, renew the hot water and leave longer. A quick recipe is to mix a can of evaporated milk with the same can full of hot water and a tablespoon of yoghurt. Pour into pots and proceed as above. Store in cold place.

Yoghurt is used as a sweet with honey, sugar, fruit; or as an appetizer, mixed with lemon juice and salt, and any or all of these: chopped cucumber, peppers, tomatoes, onion, herbs. For a thirst-quenching drink, dilute with iced water.

FRUIT 'FOOLS' are a good way of using condensed milks—both the whole and skimmed milk types. The skimmed condensed milk is cheaper and can be enriched by creaming it with 1 oz of margarine or butter.

The condensed milk is mixed with raw, stewed or canned fruit, either rubbed through a *nylon* sieve, or left whole. Some lemon or orange juice improves the texture. The fruit acid thickens the milk by clotting it and a creamy mixture is produced.

Creamy Fruit Salad. Three tins (16 oz) of fruits are put in a serving basin, with some lemon or orange juice and a tin of condensed milk stirred in. Leave to stand in a cool place for one hour. Decorate the top with fruit or grated chocolate and cream.

Instant Custard. A tin of evaporated milk is mixed with enough orange juice to thicken. Serve at once as the mixture will curdle after 2 hours.

Gooseberry Fool. Canned or stewed gooseberries are drained, then rubbed through a nylon sieve, and mixed with enough condensed milk to make a thick creamy mixture. More juice is added as needed.

Colour pale green with a skewer dipped into green colouring and pour into glass dishes.

SOUSE: this is a good way of using the bony meats, such as spareribs, pigs feet (split in half) and oxtail etc. The meat and bones are first steeped for 24 hours, then simmered for three hours in a mixture of one cup vinegar or lemon juice to 3 cups water to 2 lb meat. The liquor is strained, cooled fat removed, and then mixed with more vinegar or lemon juice (or lime juice), pickling spice, salt, and crushed black pepper corns. The meat and bones are put in a deep dish and left to steep in the liquor for 3 to 4 days in a cold place. Serve with slices of lemon and cucumber.

SWEET–SOUR. After steeping and simmering the meat and bones as in the previous recipe, the hot liquor is made into a sweet-sour sauce by adding 1 to 2 tablespoons of honey, and/or molasses, chopped garlic, onion, peppers, and ½ pint of tomato juice blended with a level tablespoon of cornflour to thicken.

The meat and bones are simmered in this sauce for another ¼ hour and served with rice or noodles. Instead of tomato juice, pineapple or orange juice can be used.

CURRY SPREAD

2 oz margarine
2 heaped tablespoons soya flour
3 heaped tablespoons curry powder
1 teaspoon yeast or meat extract
1 teaspoon sugar
1 tablespoon peanut butter or coconut cream
½ lemon, grated rind and juice

Beat all ingredients together to a creamy paste, adding more sugar, curry, or lemon juice to taste. Taste with small pieces of bread.

Use for sandwiches, with dry toast, or hot buttered toast. Serve with a glass of tomato juice, watercress. Supplies about 45 mg iron for 6 oz weight. Keeps well.

GINGERBREAD

Dry mix. Sieve on to paper: 8 oz plain flour, 2 oz sugar, 2 rounded teaspoons of ground ginger, 1 flat teaspoon of bicarbonate of soda.

Liquid mix. Collect in a saucepan: 4 oz softened margarine, 6 oz black treacle, 2 oz syrup, 2 eggs unbeaten, ¼ pt milk.

Combine the two mixes in the saucepan and beat well for one minute. The cake is baked in a paper lined tin 6 to 7 in across, at 325°F/2/150°C on middle shelf for 1½ hours. Cool in the tin for ¼ hour, then turn out on to a rack.

SESAME SEEDS

Tahini is a creamy savoury sauce made by pounding sesame seeds to a paste (which can be bought ready made), then with garlic, salt and plenty of chopped parsley. Thin to a cream with lemon juice and water.

It is eaten with bread, dry toast, poured over salads, cooked vegetables, and pulses. 1 oz supplies about 230 mg calcium.

COD'S ROE

Cod's roe, raw, must be simmered in salted water, under boiling point until firm. It can be bought cooked. The cooked roe, sliced ¼ inch thick, dipped in a batter of flour and water, is fried in oil.

The raw roe can also be soused, i.e. cooked in vinegar and water, half and half, with pickling spices, and some onion.

Taramasalata is made with smoked cod's roe. Four ounces of the smoked roe are pounded or liquidized with a slice of bread soaked in 1 tablespoon of milk and ¼ pt of oil is added slowly. Lemon juice and black pepper, milled, are added to taste. Serve with toast, potato crisps.

All three recipes retain a good deal of the thiamin in the roe.

YEAST

Yeasted pancakes. Use a 10 oz can to measure (e.g. a condensed milk can). Make a batter with one can each of flour and water or milk. The flour can be white, wholemeal, or a mixture of white and wholemeal, maizemeal or millet or oatmeal. Sprinkle in 1 teaspoonful of dried yeast (or ½ oz fresh) and a teaspoon of sugar. Leave the batter in a warm place or in the sun, for about ½ hour till frothy. Add a pinch of salt and thin down to a pouring batter with water. Fry tablespoonfuls of the batter in a little oil in a frying pan or on an oiled griddle. Turn over to brown both sides.

Nutritive value and texture are improved by mixing in an egg yolk, and folding in the stiffly-beaten white; also by stirring in a tablespoon of dried skimmed milk or soya flour. The pancakes can be served with soft or cottage cheese and sugar, lentil curry, or baked beans in tomato sauce, etc.

The addition of yeast would improve the texture and food value of such traditional pancakes as tortillas, chapattis, dampers, etc.

EASY-MAKE BREAD. Use a 10 oz can to measure the flours and water (e.g. a condensed milk can). It is convenient to mix the batter and the dough using a saucepan or casserole with lid.

1. *Make a batter* with one can each of flour and warm water, 2 rounded teaspoons of dried yeast (or 1 oz fresh) and 1 teaspoon of sugar. Leave in a warm place for at least 10 minutes till it is frothy.
2. *Make a soft dough* by adding to the batter 2 cans of flour, 1 teaspoon salt, 1 tablespoon soft margarine or oil, and more water if needed. Rake through with the hand then squeeze to make the dough.
3. *Pull and stretch* (*i.e. knead*) the dough with the hands, in the basin or on the table about 50 times until it feels firmer.

4. *Shape the dough*, either (a) to a roll and put into a greased can or loaf tin or (b) to a ball, then flatten to a round 1 in thick on a floured baking tin.
5. *Rise the dough*, covered with oiled polythene, by putting it in the sun or leave on the kitchen table until doubled in size (about ¾ to 1 hour).
6. *Bake in hot oven*, 400°F/6/200°C; about 45 minutes.

N.B. For better-keeping bread, rise the dough in an oiled pan with lid for about ¾ hour, then shape, cover with polythene and rise it again in a warm place for about ¾ hour, then bake.

Other ways of cooking bread. The round flat loaf, or smaller round flat baps can be cooked in a well-floured electric fry-pan, on a griddle, or in a thick saucepan with lid, on top of a stove or fire. It can be baked in a thick metal or earthenware pot with lid, buried in the ashes of a fire. Bread dough shaped into walnut-sized balls or into a roll, can be steamed Chinese-fashion and served with savoury dishes.

Breadmaking gives great scope for better nutrition at low cost. This basic recipe can be enriched by various additions.

Thiamin value is increased by using whole grain flour, or a mixture of white and whole grain meals—maizemeal, oatmeal, milletmeal; adding ½ can of wheatgerm or 1 tablespoon of Brewers yeast powder to the flour, or a can of crushed peanuts to the dough.

Iron value is increased by using wholemeal flour and wheatgerm, plus a ¼ can of soya flour, mixing in a tablespoon of black treacle and a can of chopped figs to make a sweet loaf.

Calcium and protein value is improved by mixing ¼ can of dried skimmed milk with the flour.

Fibre value is increased by mixing in one can of bran and another ½ can of water.

MUESLI. Two teacups of rolled oats and some dried fruit are soaked in milk for 4 to 5 hours, then mixed with a large, coarsely grated raw apple, the grated rind and juice of a lemon (or some orange juice), a tablespoon of condensed milk, some honey and chopped nuts. The mixture is served garnished with fruit—fresh or canned.

PEANUT FUDGE. Use the 10 oz can for measuring.

Combine ½ can each of peanut butter and liquid honey then work in enough powdered milk to make a stiff paste. Shape to a block ½ inch thick on oiled paper, and cut into cubes. N.B. If instant milk is used it must be crushed to a fine powder.

SAVOURY SPREAD. Baked beans in tomato sauce are mashed with grated cheese and margarine to make a stiff paste. The flavour can be varied by mixing in some yeast extract, or herbs. Use as a spread for bread or toast.

SAVOURY ROAST. The savoury spread in previous recipe is mixed with ¼ of its weight in bread crumbs (or bread soaked and squeezed dry) some onions fried in oil, mixed herbs, and enough milk or egg to bind to a stiff mix. The mixture is packed into a thickly greased tin and baked in a moderate oven ¼ hour. Serve hot with gravy or cold with salad.

New Vegetable Protein Foods (see p. 171). Look like dried meat, shaped in cubes or pieces. Reconstitute as directed and simmer ½ hour in the soaking water with 1 tablespoon of oil per ¼ lb dry weight. This makes them more juicy. To make balanced recipes and meals, foods supplying iron, calcium, vitamins A, B and C, together with carbohydrate, should be added. Suggestions for use:

1. Simmer the reconstituted protein in a strongly flavoured sauce.
   (*a*) Curry sauce, made by frying onions and garlic, in margarine with curry powder and flour, adding tomato, yeast extract and water and peanut butter and sugar to taste. Serve with rice, and relishes of dried fish, sliced peppers and onion, grated carrot and coconut. These foods provide a balanced meal.
   (*b*) Goulash sauce, onions, cooked with paprika, tomatoes, peppers, oil, thickened with flour, flavoured with yeast extract, served with noodles.
   (*c*) Sweet–sour sauce, made with orange juice or tomato juice, served with fried rice.
   (*d*) Oxtail soup, canned or in powder form makes a savoury sauce. Simmer the protein food in this with cooked carrots, serve with potatoes and greens.
2. The reconstituted protein can be cooked with rice, margarine, onions, yeast extract to make a pilaff; serve with salad.
3. Mix the protein into a yeasted flour dough, steamed to make a savoury pudding.

   Savoury pudding. Use a 10 oz can to measure.
   (*a*) Reconstitute 1 can of vegetable protein.
   (*b*) Make up a yeast liquid with 1 can warm water, 1 rounded teaspoon each of dried yeast, sugar, and flour. Leave till frothy.
   (*c*) Collect in a basin, 3 cans plain flour, ½ can suet or oil, some chopped onion, mixed herbs, 1 egg unbeaten.
   (*d*) Add yeast liquid and drained protein to basin and mix to a soft dough. Half-fill greased basins or cans, cover with polythene, leave to rise double, then steam for ½–2 hours; serve with savoury gravy or sauce, with carrots, turnips and greens.

MEAT OFFALS

These include heart, liver, kidney, melt (spleen), lungs, etc. They are cheap and good sources of protein, iron, B-vitamins, and some retinol, but are low in fat. Most countries have recipes using them. The mixed offals, together with fat bacon or pork, and onion are coarsely minced or chopped, mixed with cereal herbs and seasoning, then baked, steamed, boiled, or fried. Typical recipes are Gayettes (France), Haggis (Scotland), and the following British recipe:

FAGGOTS OR SAVOURY DUCKS

Collect in a basin the minced or chopped mixture of raw offals, $\frac{1}{4}$ of their weight in fat bacon or pork, onions, and stale bread. The bread can be dipped in water then roughly crumbled. Add sage or mixed herbs (fresh or dried), salt, and pepper to taste. Shape the mix into balls (100 g/4 oz), roll in cereal, e.g. oatmeal, or in crumbs, pack into a baking tin, and pour a tablespoon of oil over them. Bake in hot oven about 30–40 minutes. Serve hot or cold.

Variation: use finely chopped garlic and fresh ginger instead of herbs.

NOODLES

These make good use of flour as a cheap source of energy and protein and are quickly cooked by boiling for 10 minutes. They are made by mixing a cup of cold water, 1 teaspoonful salt, an egg (or $\frac{1}{2}$ cup soya flour) with as much flour as can be worked in to make a very stiff dough. Roll out as thinly as possible on an oiled board, leave a few minutes, then cut thin strips with a sharp knife, scissors, or use a machine (a good purchase for a women's co-operative). For fresh noodles dry the strips on a towel or kitchen paper for an hour, then cook in boiling water for 10 minutes. Fresh noodles can be frozen too. For dry noodles, lay the strips on a wire tray or over a line of string until dry. Store in airtight container, boil about 15 minutes. The boiled, drained noodles can be made into a tasty dish by adding any chopped cooked meat, fish, or grated cheese, with tomatoes, onion, and peppers. Serve hot with some soy sauce.

BEAN SPROUTS

In the Chinese cuisine they are always served mixed into such dishes as fried rice, chop suey, and the many savoury mixtures of thinly-sliced meat, fish, cooked with vegetables, noodles, or rice. Bean sprouts can be added raw to the cooked dish and then heated through for 5 minutes, or they are mixed in after being quickly fried in hot oil (which keeps most of the vitamin C), and then served at once. The sprouts can be eaten raw as a salad, mixed with finely-sliced onion, peppers, chopped celery, tomato, with oil and vinegar.

Bibliography

ELEMENTARY

AYKROYD, W. R. (1964) *Food for Man*, Pergamon Press, Oxford

BENDER, A. E. (1975) *The Facts of Food*, Oxford University Press

BROWN, A. M. (1966) *Practical Nutrition for Nurses*, Heinemann, London

CAMERON, A. G. (1964) *Food and its Functions*, E. Arnold, London

DURRANT, M. (1977) *Eat well & Keep Healthy*, MacDonald, London

HOUGHTON, A. A., LUND, N. A., TAYLOR, A. M. (1964) *The Food We Live On*, Allman, Mills, & Boon, London

HUTCHINSON, R. C. (1960) *Food for Longer Living*, Melbourne University Press

LEVERTON, R. M. (1960) *Food Becomes You*, Iowa State University Press

MATTHEWS, W. and WELLS, D. (1964) *First Book of Nutrition*, Flour Advisory Bureau, London

—— —— (1976) *Second Book of Nutrition*, Flour Advisory Bureau, London

MINISTRY OF AGRICULTURE, FISHERIES & FOOD (1976) *Manual of Nutrition*, HMSO, London

MOTTRAM, V. H. (1978 revised) *Human Nutrition*, E. Arnold, London

NUFFIELD ADVANCED CHEMISTRY (1971) *Food Science: A Special Study*, Longman, Harlow, Essex and Penguin, Harmondsworth

RICKETTS, E. (1966) *Food, Health, and You*, Macmillan, London

TAYLOR, T. G. (1978) *Principles of Human Nutrition*, E. Arnold, London

INTERMEDIATE

ALBANESE, A. A. (1963) *Newer Methods of Nutritional Biochemistry*, Vols. I–V. Acedemic Press, New York

DAVIDSON, SIR S., PASSMORE, R., BROCK, J. F. and TRUSWELL, A. S. (1978) *Human Nutrition and Dietetics*, Churchill Livingstone, Edinburgh

GYORGY, P. and BURGESS, A. (1965) *Protecting the Pre-School Child*, Tavistock Press, London

—— PEARSON, W. N., SEBRELL, W. H. and HARRIS, R. S. (1967–72) *The Vitamins*, Vols. I–VII, Academic Press, New York

PIKE, M. and BROWN, M. L. (1973) *Nutrition, an Integrated Approach*, J. Wiley & Sons, New York

SINCLAIR, H. M. and HOLLINGSWORTH, D. F. (1969) *Hutchison's Food and the Principles of Dietetics*, (12th ed.), E. Arnold, London

SOCIAL NUTRITION

BARKER, T. C., MCKENZIE, J. C., and YUDKIN, J. (1966) *Our Changing Fare*, MacGibbon, St. Albans, Herts

BORGSTROM, G. (1973) *World Food Resources*, International Textbook Co. Ltd, London

BRILLAT-SAVARIN, J. A. (1970) *Philosopher in the Kitchen*, Penguin, Harmondsworth

BROTHWELL, D. and BROTHWELL, P. (1969) *Food in Antiquity*, Thames & Hudson, London

BURNETT, J. (1966) *Plenty and Want*, Nelson, Sunbury-on-Thames, Middx.

DARBY, W. J., GHALIOUNGUI, P., and GRIVETTI, L. (1977) *Food: The Gift of Osiris*, Academic Press, New York

DEAN, R. F. A. and BURGESS, A. (1962) *Malnutrition and Food Habits*, Tavistock Press, London

DRUMMOND, J. C. and WILBRAHAM, A. (1958) *The Englishman's Food*, Cape, London

EHRLICH, P. R. and EHRLICH, A. H. (1970) *Population Resources and Environment*, W. H. Freeman, London

FORSTER, B. (1975) *European Diet from Pre-Industrial to Modern Times*, Harper & Row, New York

GELFAND, M. (1971) *Diet and Tradition in African Culture*, Churchill Livingstone, Edinburgh

HARTLEY, D. (1975) *Food in England*, Macdonald, London

MCCORMACK, A. (1963) *The Population Explosion and World Hunger*, Burns & Oates, London

MCKENZIE, A. (1970) *The Hungry World*, Faber & Faber, London

ODDY, D. and MILLER, D. S. (ed.) (1976) *The Making of the Modern British Diet*, Croom Helm, London

PYKE, M. (1968) *Food and Society*, John Murray, London

STEWART, K. (1975) *Cooking and Eating*, MacGibbon, St. Albans, Herts

TANNAHILL, R. (1973) *Food in History*, Paladin, St. Albans, Herts

WALCHER, D. N., KRETCHMER, N., and BARNETT, H. L. (1976) *Food, Man, and Society*, Plenum Press, New York

WARD, B. and DUBOS, R. J. (1972) *Only One Earth*, André Deutsch, London

YUDKIN, J. and MCKENZIE, J. C. (1964) *Changing Food Habits*, McGibbon, St. Albans, Herts

SPECIAL TOPICS

ALTSCHUL, A. M. (1965) *Proteins: their Chemistry and Politics*, Chapman Hall, London

AMERINE, M. A., PANGBORN, E. B., and ROESLER, E. B. (1966) *Principles of Sensory Evaluation of Food*, Academic Press, New York

AYKROYD, W. R. and DOUGHTY, J. (1970) *Wheat in Human Nutrition*, FAO, Rome

—— —— (1970) *Legumes in human Nutrition*, FAO, Rome

AYRES, J. C., KRAFT, A. A., SNYDER, H. E., and WALKER, H. W. (1969) *Chemical and Biological Hazards in Foods*, Hafner Publication Co., New York

BENDER A. E. (1978) *Food Processing and Nutrition*, Academic Press, London

BIRCH, G. G., CAMERON, A. G., and SPENCER, M. (1972) *Food Science*, Pergamon Press, Oxford

—— (ed.) (1976) *Food from Waste*, Applied Science Publishers, Barking, Essex

BRAVERMAN, J. B. S. (1976) *Introduction to the Biochemistry of Foods*, Elsevier Publishing Co., Amsterdam

CAKEBREAD, S. H. (1975) *Sugar and Chocolate Confectionary (The Value of Food Series)*, Oxford University Press

CHRISTIE, A. M. (1975) *Food Hygiene and Food Hazards*, Faber & Faber, London

COPSON, D. A. (1975) *Microwave Heating*, Avi Publishing Co., Westport, CT

COX, P. M. (1968) *Deep Freezing*, Faber & Faber, London

DESROSIER, N. W. (1970) *The Technology of Food Preservation*, Avi Publishing Co., Westport, CT

DUCKWORTH, R. (1966) *Fruit and Vegetables*, Pergamon Press, Oxford

DUMONT, R. and ROSER, B. (1969) *The Hungry Future*, Deutsch, London

DURNIN, J. and PASSMORE, R. (1967) *Energy Work and Leisure*, Heinemann, London

EDHOLM, D. G. (1967) *Biology of Work*, World University Library, London

FOMON, S. J. (1974) *Infant Nutrition*, W. B. Saunders, Philadelphia

FRANCIS, D. E. M. (1974) *Diets for Sick Children*, Blackwell, Oxford

GEORGE, S. (1976) *How the Other Half Dies*, Penguin, Harmondsworth

GERRARD, F. (1971) *Meat Technology*, Leonard Hill Press, London

GOODHART, R. S. and SHILS, M. E. (1974) *Modern Nutrition in Health and Disease*, Lea & Febiger, Philadelphia

GRISWOLD, R. M. (1962) *The Experimental Study of Foods*, Houghton & Mifflin, New York

GUNTER, M. (1973) *Infant Feeding*, Penguin, Harmondsworth

HAWTHORN, J. and LEITCH, J. M. (1962) *Recent Advances in Food Science*, Vols. I and II. Butterworth, London

HESKIN, N. A. M., HENDERSON, H. M. and TOWNSEND, R. J. (1971) *Biochemistry of Foods*, Academic Press, New York

HOBBS, B. C. (1968) *Food Poisoning and Food Hygiene*, E. Arnold, London

——— and CHRISTIAN, J. H. B. (ed.) (1973) *The Microbiological Safety of Food*, Academic Press, London

HOWARD, A. N. and MCLEAN BAIRD, I. (ed.) (1973) *Nutritional Deficiencies in Modern Society*, Newman Books, London

LAPPE, F. M. (1975) *Diet for a Small Planet*, Ballantine, New York

LATHAM, M. (1965) *Human Nutrition in Tropical Africa*, FAO, Rome

LAWRIE, R. A. (1974) *Meat Science*, Pergamon Press, Oxford

LOWE, B. (1955) *Experimental Cookery*, J. Wiley & Sons, New York

MARKS, H. (1968) *The Vitamins in Health and Disease*, Churchill, Livingstone, Edinburgh

MARTZ, W. and CORNATZER, W. E. (1971) *Newer Trace Elements in Nutrition*, M. Dekker, New York

MASEFIELD, G. B., WALLIS, M., HARRISON, S. G., and NICHOLSON, B. E. (1971) *The Oxford Book of Food Plants*, Oxford University Press

MCCOLLUM, ELMER V. (1975) *History of Nutrition*, Houghton Mifflin, Boston

MCLAREN, D. S. and BURMAN, D. (1976) *Textbook of Paediatric Nutrition*, Churchill Livingstone, Edinburgh

MCLEAN BAIRD, I. and HOWARD, A. N. (eds.) (1969) *Obesity: Medical and Scientific Aspects*, Longman, New York

MEYER, L. H. (1969) *Food Chemistry*, Van Nostrand & Reinhold, New York

PAUL, P. C. and PALMER, H. H. (1972) *Food Theory and Applications*, J. Wiley & Sons, New York

PEARSON, D. (1970) *The Chemical Analysis of Foods*, (6th edn.), Churchill Livingstone, Edinburgh

PIRIE, N. W. (1969) *Food Resources, Conventional and Novel*, Penguin, Harmondsworth

PORTER, J. W. G. (1975) *Milk and Dairy Foods*, (The Value of Food Series). Oxford University Press

PYKE, M. (1964) *Food Science and Technology*, John Murray, London

——— (1950) *Industrial Nutrition*, John Murray, London

——— (1974) *Catering Science and Technology*, John Murray, London

ROBINSON, C. H. (1972) *Proudfit-Robinson's Normal and Therapeutic Nutrition*, (13th edn.). Macmillan, London

SCADE, J. (1975) *Cereals* (The Value of Food Series). Oxford University Press

SPAULL, H. (1961) *The World Unites Against Want*, Barrie & Rockliff, London

SPICER, A. (ed.) (1975) *Bread*, Applied Science Publishers, Barking, Essex

STADELMAN, W. J. and COTTERILL, O. J. (ed.) (1973) *Egg Science and Technology*, Avi, Westport, CT

WELBOURN, H. (1963) *Nutrition in Tropical Countries*, Oxford University Press

WOOLLEN, A. (ed.) (1969) *The Food Industries Manual*, Leonard Hill Press, London

WYNNE-TYSON, J. (1975) *Food for a Future*, Davis-Poynter, London

OFFICIAL PUBLICATIONS. FOOD TABLES. DICTIONARIES.

BENDER, A. E. (1974) *Dictionary of Nutrition and Food Technology* (4th edn), Butterworth, London

CAMERON, M. and HOFVANDER, Y. (1971) *Manual on Feeding Infants and Young Children*, Protein Advisory Group (PAG) of the UN System

DAVIS, J. G. (1955) *Dictionary of Dairying*, Leonard Hill Press, London

DEPARTMENT OF EDUCATION AND SCIENCE REPORT (1973) *Nutrition in Schools*, HMSO, London

DEPARTMENT OF HEALTH AND SOCIAL SECURITY (1969) *Family Expenditure Survey Handbook*, HMSO, London

——— (1969) *Recommended Intakes of Nutrients for UK*, HMSO, London

——— (1970) *Food Hygiene Regulations*, HMSO, London

——— (1972) *Clean Catering* (4th edn.), HMSO, London

——— (1972) *A Nutrition Survey of the Elderly. Reports on Health and Social Subjects No. 3*, HMSO, London

——— (1974) *Present Day Practice in Infant Feeding*, HMSO, London

——— (1975) *Catering in Homes for Elderly People*, HMSO, London

——— (1975) *Nutrition and Modified Diets*, HMSO, London

——— (1975) *Nutrition Survey of Pre-School Children*, HMSO, London

——— (1976) *Diet and Coronary Heart Disease*, HMSO, London

FAO, (1970) *Report of the Second World Food Congress*, FAO, Rome.

JELLIFFE, D. B. (1966) *Assessment of the Nutritional Status of the Community*, WHO, Geneva

JOINT FAO/WHO COMMITTEE REPORT SERIES NO. 522 (1973) *Energy and Protein Requirements*, FAO, Rome

MCCANCE, R. A. and WIDDOWSON, E. M. (1978) *Composition of Foods*, Revised by A. A. Paul and D. A. T. Southgate). HMSO, London

MINISTRY OF AGRICULTURE, FISHERIES AND FOOD (1974) *Second Report on Bread and Flour by the Food Standards Committee*, HMSO, London

——— *Household Food Consumption and Expenditure*, (Produced annually) HMSO, London

PLATT, B. S. (1945) *Tables of Representative Values of Foods Used in Tropical Countries. Special Report Series No. 302*. HMSO, London

COOKERY BOOKS

CLARK, M. (1970) *Teaching Cookery*, Pergamon Press, Oxford

DAVIES, L. (1974) *Easy cooking for one or two*, Penguin, Harmondsworth

—— (1975) *Easy cooking for three or more*, Penguin, Harmondsworth

DAVIS, A. (1963) *Lets Cook it Right*, Allen & Unwin, London

ELLISON, A. (1966) *The Great Scandinavian Cook Book*, Allen & Unwin, London

EWALD, E. B. (1975) *Recipes for a Small Planet*, Ballantine, New York

FISHER, P. F. (1963) *Lets Cook With Yeast*, Allman, Mills & Boon, London

—— (1969) *500 Vegetarian Recipes*, Hamlyn, London

GARDINER, M. (1962) *Food Preparation*, Constable, London

GLEW, G. (ed.) (1973) *Cook Freeze Catering, Introduction to its Technology*, Faber & Faber, London

MILLROSS, J., SPEHT, A., HOLDSWORTH, K., and GLEW, G. (1973) *The Utilisation of the Cook-Freeze Catering System for School Meals*, University of Leeds

NILSON, B. (1964) *Cooking for Special Diets*, Penguin, Harmondsworth

—— (1967) *The Penguin Cook Book*, Penguin, Harmondsworth

EDUCATION AND TEACHING NUTRITION

DE ESQUEF, L. (1975) *Food and Nutrition Education in The Primary School*, FAO, Rome

FRIDTHJOF, J. (1964) *Encouraging the Use of Protein-Rich Foods*, FAO, Rome

HOLMES, A. C. (1968) *Visual Aids in Nutrition Education*, FAO, Rome

MARTIN, E. A. (1963) *Nutrition Education in Action (A Guide for Teachers)*, Holt, Reinhart & Winston, New York

RITCHIE, J. (1965) *Learning Better Nutrition*, FAO, Rome

FAO—Food and Agriculture Organisation, United Nations, Rome

WHO—World Health Organisation, Geneva

HMSO—Her Majesty's Stationery Office, London

Index